NUNEATON & BEDWORTH

COAL, STONE, CLAY & IRON

NUNEATON & BEDWORTH

COAL, STONE, CLAY & IRON

FORGOTTEN INDUSTRIES IN NORTH WARWICKSHIRE

PETER LEE

AMBERLEY

First published 2011

Amberley Publishing
The Hill, Stroud
Gloucestershire, GL5 4EP

www.amberleybooks.com

British Library Cataloguing in Publication Data.
A catalogue record for this book is available from the British Library.

ISBN 978 1 84868 970 1

Typesetting and Origination by Amberley Publishing.
Printed in Great Britain.

PREFACE

For the first fifty years of the twentieth century, the Warwickshire towns of Nuneaton and Bedworth enjoyed full employment. The rapid increase in population was due to the availability of jobs and the need for labour. The range of work available was diverse. By far the largest employers were the coal and brick industries. These employed tens of thousands of hands. While the men worked down the pit or in the brickyards, their wives could find employment in the textile industry, mills and light industrial factories found in abundance throughout the district. Additionally, there was a need to move bulky freight. That too created a large number of employment opportunities on the railways. Industries such as mining and brick-making, machinery manufacture, iron founding, general engineering and local coal distribution and cartage were big employers.

This part of North Warwickshire had never had it so good. Since the 1960s, we have seen almost the entire loss of these once important employers. In employment terms, the twin towns of Nuneaton and Bedworth have gone into terminal decline; they are now classed as the most deprived in the county of Warwickshire. The heavy industries, with their attendant job opportunities, have never been replaced and almost certainly never will be. The borough has the highest unemployment in Warwickshire, and in 2010 48.9 per cent of the working people of the twin towns (with a combined population of 120,000) commute outside of the borough for work. There is one working coal mine left in the north of the county – Daw Mill – and roofing tiles are also still manufactured in Bedworth. That is all. It is a sobering thought that almost all of the vast industrial activity recorded in this book has vanished, and is most likely gone forever.

From the beginning of the seventeenth century, the area we know now as the Borough of Nuneaton and Bedworth was riddled with coal workings and brick kilns. Wherever coal was found, good brick clay was found alongside. As the clay was penetrated while reaching the coal-seams, it made sense to construct one or more small beehive kilns to produce bricks for the pithead buildings, and the surplus could be sold to the local building trade. As you travel around the local countryside, you may find the remains of old buildings long since demolished and, impressed in the frogs of the bricks (the depressed central part) scattered about, the words 'Durex', 'Hardy', 'Haunchwood', 'Moorwood', and 'Stanley Brothers' among others. Such are the remnants of this mighty industry now obliterated from the landscape.

In addition to clay, there were fine beds of building stone. Attleborough freestone was at one time a popular building material for churches, mansions and fine public buildings, whereas Hartshill Quartzite (often referred to as granite because of its very durable qualities) was a hard and dense material that was used for road metalling and, with the advent of railways, railway ballast. These materials could not be bettered anywhere in England. They were in great demand throughout the country. The quarrying industry, important commercially as it was, did not employ the same numbers of workers as the coal and brick industries, but it was important nevertheless. Stone was transported by rail in vast quantities – a lot of customers were railway companies – or by canal where the rail sidings did not reach. Later, of course, the stone was delivered by road.

An offshoot of the coal industry was ironstone mining. Nodular ironstone was found among the coal-bearing strata and it was thought worthwhile for a time to extract it. In fact, some shafts were sunk for this specific purpose. In Bedworth, great expense was outlaid to build blast furnaces to process this ironstone into iron. However, these were very short-lived affairs. Ironstone is an insignificant contribution to our story, but it is worth mentioning for the sake of completeness.

Local production of coal consisted mostly of house coal for domestic use, for power generation, and it had certain characteristics suitable for the process industries, particularly brewing.

The aim of this book is to provide a summary of the known enterprises associated with the industries under review as far back as it has been possible to go. As you would expect, at first they were small-scale concerns, often operated by one man, or partnerships on land leased from local landowners. The landowners received royalties on every ton of coal dug. Occasionally, landowners would also work the coal-seams; the Newdigate family particularly did so at times. But mostly they left it to traders to carry on the business, content to take the royalties, avoiding the risk and the hard work. The middle of the nineteenth century saw some changes to the way the management of the various businesses was organised. The concept of limited liability companies enabled entrepreneurs to take more chances. People travelling from a wider area came into this district knowing its mineral potential. They were often skilled mining engineers with expert knowledge. The advent of canals and then the railways revolutionised distribution, dramatically reducing the cost of the product at the point of delivery.

The Victorian period was a time of great change in engineering. Improved mining, brick-making and quarrying equipment aided by better metallurgy meant that more durable and productive machinery could be built. All these factors coming together meant that the small-scale primitive businesses of the early 1800s, which had restricted their various outputs to a local area, had coalesced into large limited liability companies by 1900. They were sending their products all over England, and in some cases – particularly with regard to fine-quality terracotta products and blue engineering bricks – all over the civilised world. There are not many great buildings in London that do not have Nuneaton and Bedworth terracotta products in them.

You will see numerous references to shafts, pits, mines and collieries in this book. To clarify this: an individual shaft became either a mine or a colliery when it was grouped with others into a single trading entity. So, the usual minimum for a colliery is two shafts, one being provided as an emergency escape route, but very often there are numerous shafts used for different purposes. Additional shafts were provided for pumping or ventilation purposes or in order to access remote parts of the colliery and different seams at different depths.

'Pits' could be classified as single shafts, but the term was also used colloquially to refer to a colliery. When coal mining started in the Nuneaton and Bedworth area, the seams mined were along the exposed coal measures. The coal was just below the surface of the ground, or outcropped. In order to access the coal, a series of what are termed 'bell pits' was sunk. These were so called because they were dug in a bell shape at the bottom. Primitive mining methods prevented long tunnels being dug underground through lack of ventilation; so, in order to carry on working the coal, another pit had to be sunk just a few yards away. In this way, the whole line of exposed coal measures between Bedworth and Nuneaton Common was riddled with shafts that cut a swathe through the countryside.

In order to prevent the sides of the shafts from caving in, a crude wooden framework was built which braced the sides. It was common practice when one bell pit was worked out to remove the support frame, and the gin wheel or whimsy, then transfer it to the next diggings. In this way as the bell pits moved along the seam they left a crumbling series of collapsing holes in their wake. One could imagine it was a chancy business in the seventeenth and early eighteenth century taking a walk alone across the coal pit fields – you might suddenly disappear down an unfenced pit shaft never to be seen again, your cries unheard.

Bell pits were in effect a series of single shafts that may never have been worked collectively as a group so would just have been referred to as a pit. Some of the early pits mentioned in this book may just be bell pits that had a productive life long enough to have acquired a name, only to be abandoned soon afterwards.

The philosophy behind the book is to concentrate on the extractive businesses themselves. There were many firms closely associated with these: iron founding particularly, railways and also canals. But these are large subjects in their own right, so I have only mentioned them in connection with their direct ownership by the colliery, brickworks or quarry concerned.

I should mention that the principal railway lines connected to the many undertakings under review were the London & North Western Railway's (LNWR hereafter) Coventry to Nuneaton branch, and the Trent Valley Railway, or the Midland Railway's (MR hereafter) Leicester to Birmingham line. These principal rail routes are still used today, but you will find it difficult to identify the site of many interchange sidings that used to bring traffic on to the system in the old days. Both the LNWR and the MR became major components of the London Midland & Scottish Railway in 1923, which then became part of the nationalised British Railways in 1948.

Finally this book is about people – good, hard-working people who lived and strived in the enterprises under review. It is dedicated to their memory.

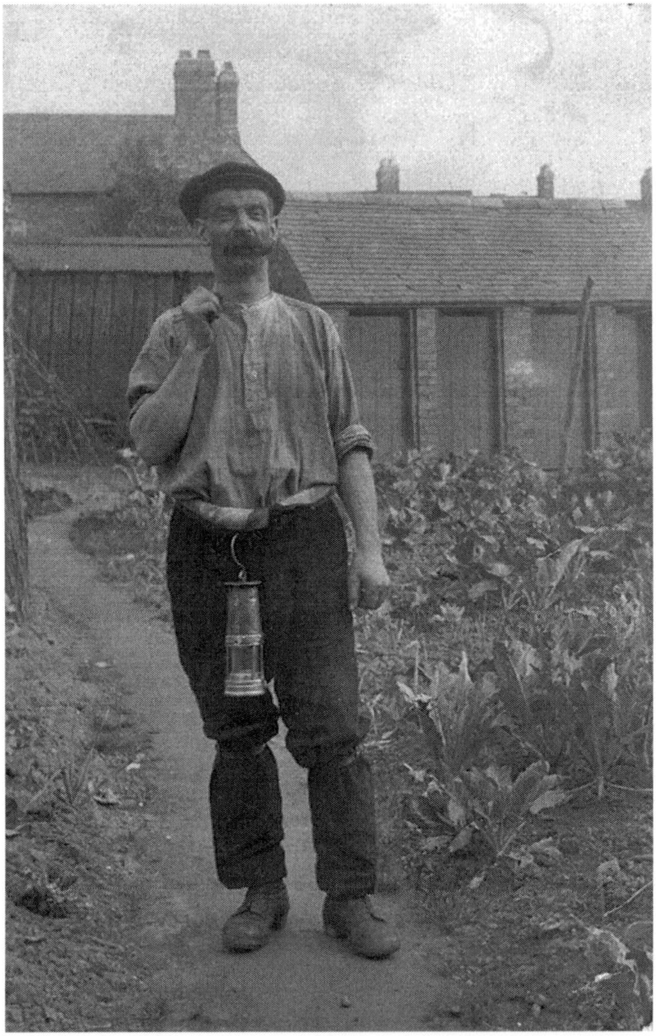

A typical local collier in his work clothes. This photo was found in a local second-hand shop and may illustrate a Nuneaton or Bedworth miner. At any rate, it shows the standard working clothes of a proud working man, complete with his Davy lamp. His relatively cleanly state suggests this was before he went to work. Note the vegetable patch. There was less emphasis on flowers in miners' gardens, as they depended on their modest plots to supplement their diets. The doorways of a row of privies complete the scene. (Jean Lapworth)

ACKNOWLEDGEMENTS & SOURCES

My grateful thanks to the following contributors of photographs and general information adding to my knowledge over the years. Without them this book could not have been written.

Steve Adnett, A. L. P. Ambrose, Thomas Anney, Clive Ball, Peter Bayly, Maurice Billington, Reg Bull, Anthony Case, Alan Cook, *Coventry Evening Telegraph*, Alan Croshaw, Keith Draper, Geoff and Madge Edmands, Dean Elliott, Roy Etherington, Ray Fox, Albert and Lawrence Fretwell, Jim Gordon, Ian Hickman, Bill and Vic Holloway, John Jeavons, Christopher Jones, Beryl Kerby, C. Kind, Mike Kinder, Dennis Labrum, Jean Lapworth, Jim Lee, Michael J. Lee, *Nuneaton News*, John Nunn, Tony Parrott, Bea and Fred Phillips, Mrs E. Richardson, June Robinson, Reg Rowley, David Sidwell, Don Slaney, Eric Tonks, Phillip Vernon, D. M. Watson, Gerry Wisher, Eileen Worthington, the staff of Nuneaton Library, the staff of Warwickshire County Records Office, the Industrial Railway Society, the London & North Western Railway Society, the Narrow Gauge Railway Society, and the Stephenson Locomotive Society.

SOURCES

- Catalogue of plans of abandoned mines, Warwickshire, HMSO, 1929.
- H. H. Howell, *The Geology of the Warwickshire Coalfield and the Permian Rocks and Trias of the Surrounding District*, 1859.
- Industrial Railway Society, *Industrial Railways & Locomotives of Warwickshire*, 2003.
- *Memoirs of the Geological Survey of Great Britain and of the Museum of Practical Geology.*
- James Mitchell, *Report on the Employment of Children and Young Persons in the Mines of the Warwickshire and Leicestershire Coal-Fields, and on the State, Condition, and Treatment of such Children and Young Persons*, 1842.
- NCB West Midlands Area No. 4 Returns, 1960.
- Peak District Mines Historical Society Ltd: Warwickshire's Mining Industry in 1896 – A List of Coal Mines. <http://www.pdmhs.com/1896%20Lists/1896-49.htm>
- F. C. Swallow, *Coal-Mining in Warwickshire, with special reference to the use of Stanley coal-heading machines in the rapid development and working of the Nuneaton Colliery*, 1903.

INTRODUCTION

Warwickshire coal-mining started on a commercial scale in the reign of Elizabeth I. Between 1550 and 1700, coal production in the Warwickshire coalfield increased on a greater scale than anywhere else in England. Before 1500, very little coal was mined at all in the Bedworth and Nuneaton area. In 1544, the 'Proffytt of Collpitts' in Nuneaton was only £2 13*s* 4*d*.

In 1570, Crown commissioners entered Warwickshire to review the availability of coal in the recently discovered coalfields of Bedworth. They visited the new coal mine at Bedworth then surveyed the area around, looking for the furthest extent of coal in the area. As a result of their visit, tradesmen were encouraged to come to the coal district to carry on the coal trade. During the last period of Elizabeth's reign, leases that had hitherto been worth a few shillings had become far more valuable – we read of coal entrepreneurs running up thousands of pounds worth of debt to buy plant and equipment for their new mines. A large impetus for this trade came from the nearby city of Coventry, which was one of the biggest cities in England. Despite the increase in production, Coventry was always short of coal. In addition, coal was transported along the Watling Street and the Fosse Way by packhorse. Local coal was known to have been traded in Leicester, Northamptonshire, South Warwickshire and Oxfordshire. Traders entered Coventry with cloth to be sold and returned to their homes with coal bought on the local market in the city.

The coalfield covered approximately 100 square miles. The principal coal-seams were:

- Four Feet
- Two Yards
- Ryder Seam
- Slate Coal
- Seven Feet
- The Bench
- Double Coal

In the nineteenth century, there were approximately 300 coal shafts in the Nuneaton and Bedworth areas. Output in Warwickshire was:

- 1860 545,000 tons
- 1881 1,113,419 tons
- 1906 3,939,497 tons
- 1932 4,794,648 tons
- 1933 4,624,348 tons
- 1934 5,054,856 tons

Between 1850 and 1913, 550 Warwickshire miners died as a result of their employment.

By 1961, the output of saleable coal in the Warwickshire coalfield was 3.77 million tons, from coal measures 25 miles long and a maximum of 8 miles in width, formed of twenty seams of which twelve were of good workable thickness and quality. The thickest seam, known appropriately as the 'Thick coal', was an amalgamation of narrower seams, giving a thickness of 20–30 feet.

THE BRICK & TILE TRADE

By the 1850s there were eighty small brickyards operating in the Warwickshire coalfield, producing 75.5 million bricks in 1858.

CANAL DISTRIBUTION

The principal means of moving coal in the local coalfield prior to 1850 was the Coventry Canal. The local canal made important connections with the Oxford Canal at Hawkesbury and the

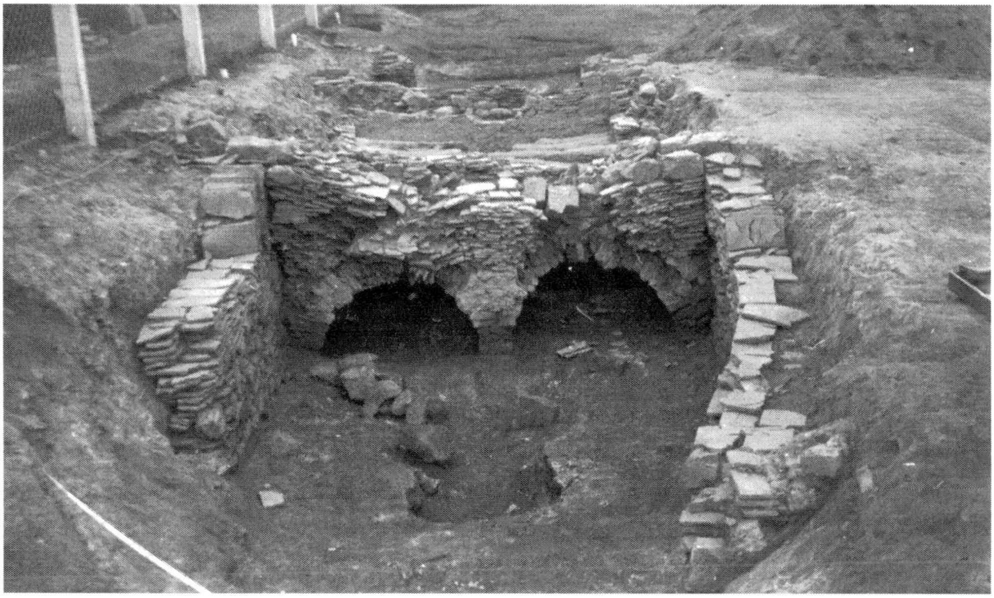

Photographed on 16 September 1967, this medieval kiln was discovered during building works near to the old Griff No. 4 colliery for the footings for a new industrial unit. It was then carefully excavated and is a fine example of an ancient tile kiln. (Geoff Edmands)

Ashby Canal near Nuneaton. The arrival of the Coventry Canal brought about a revolution in the economic transport of coal over great distances. Although the Coventry–Nuneaton railway line took some of this trade, canal transport remained an essential ingredient of coal distribution until the closure of the mines. Indeed, during the First World war, due to congestion on the railway, it was deemed quicker to send coal from the Griff and Bedworth pits to Coventry by canal than by rail. No intermediate shunting was needed. Coal could be loaded at the canal wharf and taken direct to its destination.

THE STONE TRADE

The principal types of stone produced locally were Attleborough sandstone (see the section on Attleborough Stone Quarries for a fuller description) and Hartshill Quartzite, a hard, dense material that was in great demand as a road stone. It was especially useful in macadamising roads, and in making stone blocks or setts, being dust-free for railway ballast and sewage works media. When crushed, it was used for all kinds of concrete products.

GENERAL NOTE ON THE PREPARATION OF THIS BOOK

Good photographs of many of these operations do not exist. Up until 1900, photography of the enterprises seen here was rare. It was probably thought that such scruffy, run-down establishments were not worth recording. After 1900, picture postcards became common, and some of the firms were photographed then. It is fortunate that cameras were on hand to record what they did. By the 1950s, railway enthusiasts were touring the district, taking pictures of the many industrial firms with steam locos. The focus was on the steam locomotive, but, as a result, they captured the fascinating backdrop of buildings, workings and other evidence that is valuable today. That is why there are a substantial number of rail-related pictures featured, as these were the only photographs taken.

THE BUSINESSES
(ARRANGED ALPHABETICALLY)

CHARLES ABELL HARTSHILL GRANITE QUARRIES

Owned by:

- Richard L. Alkin until about 1880
- Charles Abell until 1905
- Charles Abell Ltd until 1951
- Man-Abell Quarries Ltd from 1951
- Judkins Ltd from April 1962

The siding was fed by a half-mile tramway worked down by gravity and hauled up by rope. Moving wagons to the quarry face was carried out by horses. As late as June 1961, five horses were still used for this purpose. The horses were called Jack, Sam, Bill, Bob and Bet. Transhipment to the main-line rail system occurred at a rail wharf 2¾ miles north of Nuneaton station on the LNWR Trent Valley main line. There was a narrow-gauge connection to the Coventry Canal.

Concrete paving slabs were made on the site under the brand name 'Alexandra Paving'. Locomotive haulage was introduced in 1919. This consisted of two standard Bagnall quarry locos of virtually identical construction.

JOSHUA WATERS ADAMS (1810–1889)

Brick-maker of Chilvers Coton (and timber merchant). Married Elizabeth, one of the daughters of William Bindley. Bindley was also a brick-maker (q.v.). Adams may have taken over William Bindley's operations, after the latter's death. The brickyard was in Coton Road and appears to have been leased from the Harpur estate in 1849 for a period of fifteen years. Adams Employed thirteen men and eight boys in 1881. Elizabeth was also listed as a brick-maker in her own right, prior to her marriage.

JOHN ALEXANDER

Inventory of estate is dated 1680. Brick maker. One kiln, Wash Lane, Nuneaton. Made bricks, quarries, tiles, gutters. Mr Alexander carried on a considerable business.

GEORGE ALLEN

Listed in trade directories for 1828, 1835, 1841. Brick-maker, Abbey Street, Nuneaton (yard location unknown).

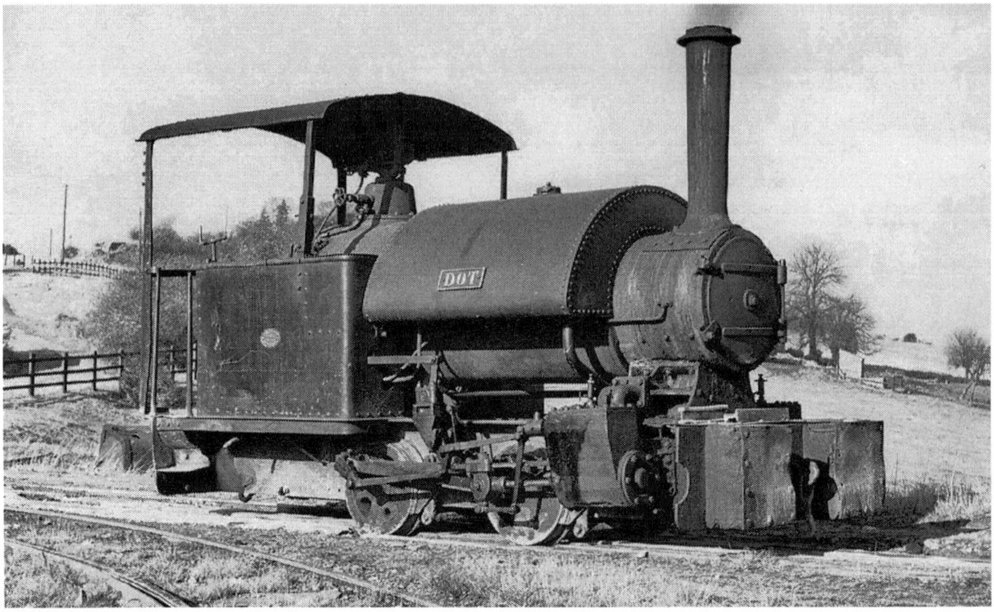

The narrow-gauge tramway at Abell's was 3-foot gauge and featured two of these little Bagnalls. Here is *Dot* standing outside the loco shed. Its partner was named *Kitty*. *Dot* was purchased new from Bagnall's in 1923. The other loco had been built in 1918 and came from the Ministry of Munitions second-hand in 1919. It was originally built for the trench railways in the First World War, but delivered as the war ended and put up for sale without being sent to France or being used in a military zone. Both engines were in use up until April 1956 and lingered on at the quarry after being withdrawn from service. *Dot* seen here was scrapped in March 1958, while its partner was scrapped early in 1960. A diesel arrived in April 1956 to replace the steamers. Originally purchased from Ruston & Hornsby's in 1942, it had worked nearby at Mancetter Quarries and had to be re-gauged from their 2 feet 8½ inches to Abell's gauge. The diesel was unnamed at Abell's. It lasted until Judkins took over in 1962, when they abandoned the railway system, and was sold to a dealer T. W. Ward of Templeborough Works, Sheffield, in either later 1964 or early 1965. The diesel's ultimate fate is unknown.

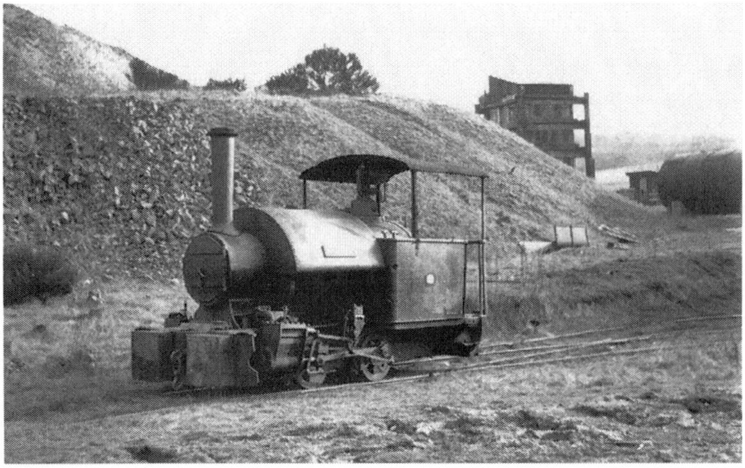

Another view of a quarry Bagnall, now withdrawn from service and pulled from the shed for the unknown photographer. It illustrates the bleak and windswept aspect of the typical quarry scene, scattered with strange buildings and industrial junk. (Author's collection)

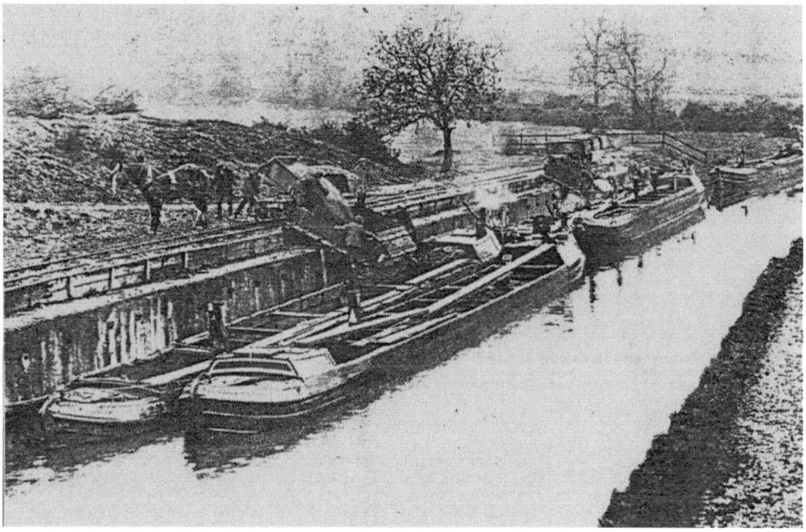

Again a poor picture but important, as it illustrates the transfer of stone from horse-drawn wagons into barges at the canal wharf. Stone was often distributed by canal for the use of turnpike trusts and in connection with road metalling. Canals sometimes reached unloading places that were not connected by rail. (Author's collection)

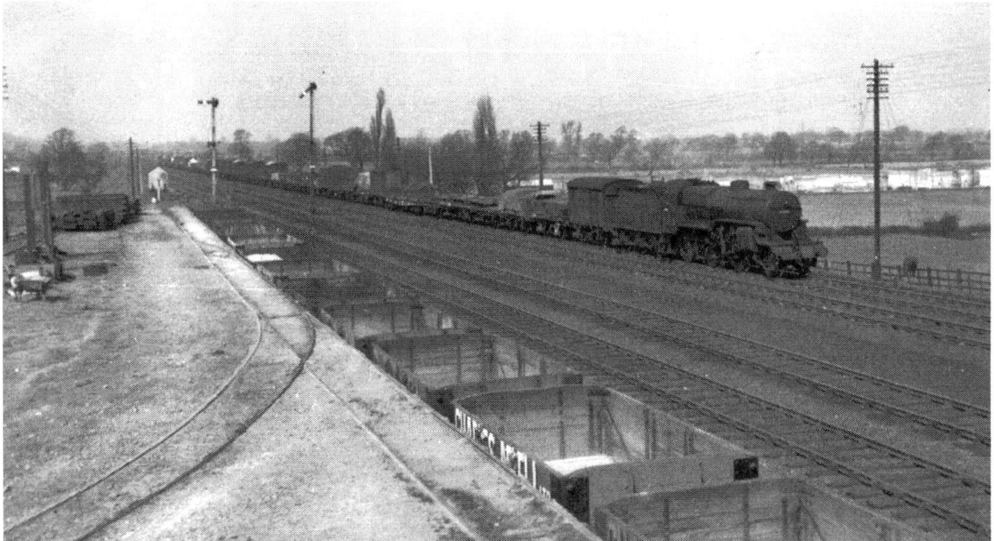

A view of Abell's rail tipping dock. The date must be when Abell's were still using their own private railway trucks, which can be seen parked in the sidings adjacent to the wharf. The quarry tramway is clearly seen with its track right on the edge of the dock. An LMS Fowler 'Crab' 2-6-0 heads south with a very mixed freight. The narrow trackwork on the left was worked by gravity downwards and hauled up by rope. Positioning of wagons on the dock was carried out by horses that were stabled in a shed just off the picture on the left. Shortly after this photo was taken, the signals on the main line were changed to upper quadrant. (R. S. Carpenter)

ANSLEY HALL COAL & IRON LTD

The colliery business commenced in 1873, and the coalfield was noted as covering 2,800 acres. Later, the colliery company worked 4,000 acres, including the Oldbury estate. Brick-making started and the pit was sunk in 1873–74. There were eleven seams 80 feet thick, which in 1873 were said to be capable of yielding 2,000 tons per day for 300 years.

Bricks had been made in the Ansley area since 1760. The clayhole was known as 'Neddy Orton's Pit'. Initially, there was a single beehive kiln, but this was soon superseded by a continuous kiln that produced bricks from pit debris. A single chamber of the kiln could hold 13,000 bricks. A Bradley Craven 'Stiff Plastic' brick-making machine was purchased in 1896, and there was a German 'Hoffmann' twelve-chamber continuous kiln in use at one time.

In 1891, 225 men were employed. In 1894, there were 184 employees producing 70,000 tons of coal per annum. By 1896, there were 228 underground and 69 surface workers. In 1913, there were 391 men producing 139,000 tons per annum. In 1939, 674 men were employed and 670 in 1945. In 1957, production was 224,600 tons of coal with 724 men.

Brick-making continued for a longer period than the colliery business, not ceasing production until December 1969. It was carried out on a substantial scale, with 3 million engineering bricks produced annually. A total of 66 million bricks were produced between 1947 and 1969.

The National Coal Board built an area workshop for concentration of engineering activities on the Ansley Hall site.

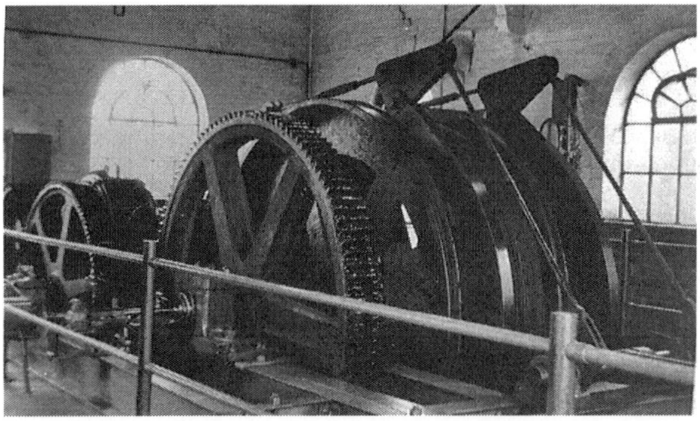

One of the winding engines at Ansley Hall Colliery. This view was probably taken in the 1950s. (Don Slaney)

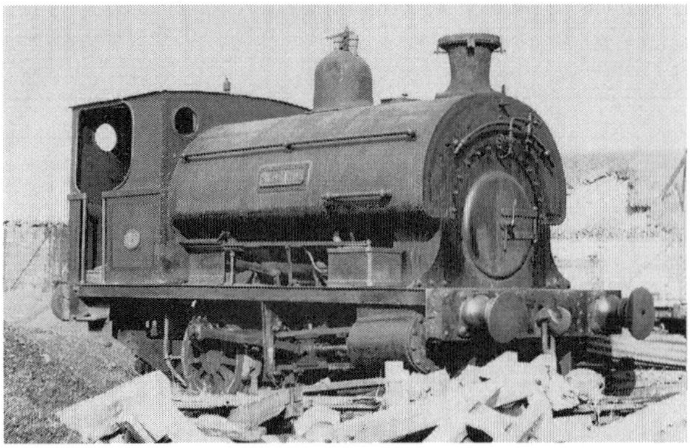

Ansley Hall was a standard Peckett works No. 991 of 1905. Purchased new, it was rebuilt with a new boiler by Andrew Barclays in 1939. After nationalisation, it had a stint at Arley Colliery between 1950 and 1952 and was then transferred to Binley Colliery in Coventry in February 1960.

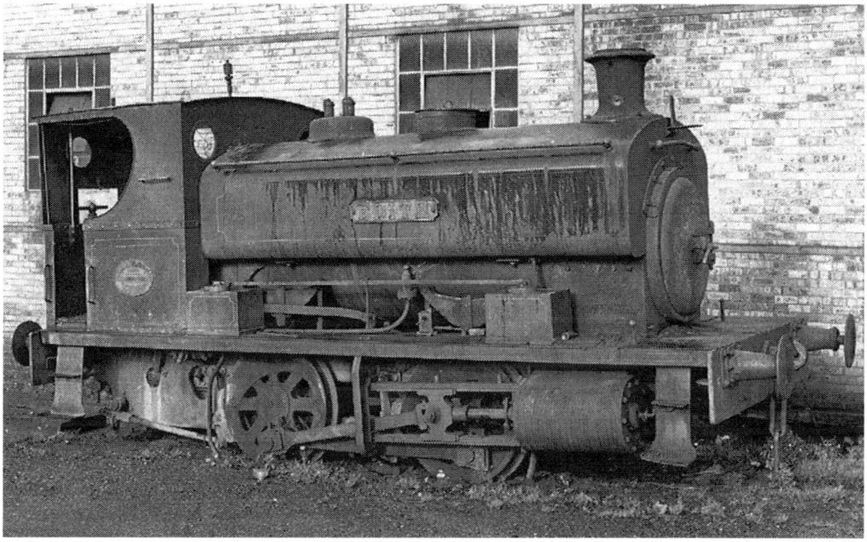

Edith was purchased new by Ansley Hall Colliery in 1913 to replace a loco by the same name that was sold to the Wigan Wagon Co. Ltd in Lancashire. The former engine had been built in 1886 and came from Hawthorn Leslie. This was one of the 'chunky' products of Andrew Barclay that provided robust shunting power from 1913 until rebuilt by Woodward & Kingsley at their Station Works, Kingsbury, between March and August 1954. It returned to Ansley Hall before being transferred to Arley Colliery two weeks before closure on 14 November 1959 of the colliery, where it is photographed looking forlorn and derelict. It was scrapped on site at Arley by George Cohen & Sons Ltd during February 1964.

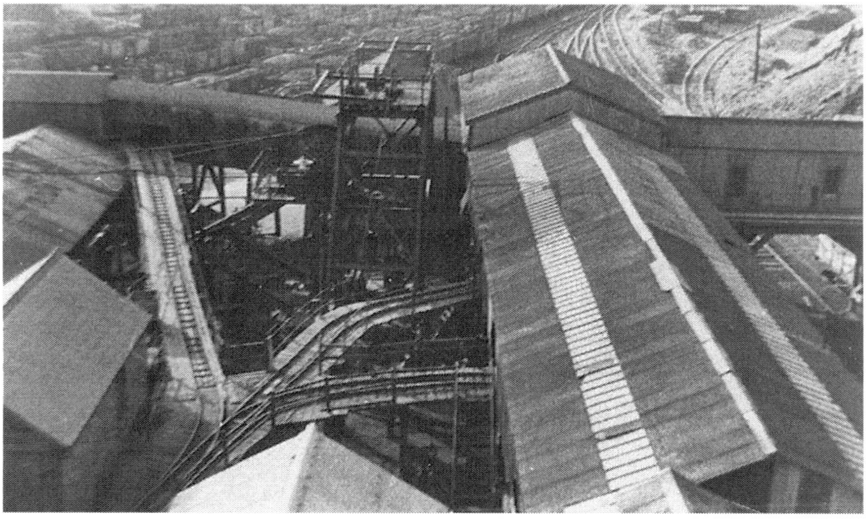

An interesting view from the pithead at Ansley Hall Colliery in 1958 – upcast shaft headgear showing the complex layout of tub tracks distributing coal around to different locations and conveyor points. Note also the large number of wagons parked in the sidings. (Don Slaney)

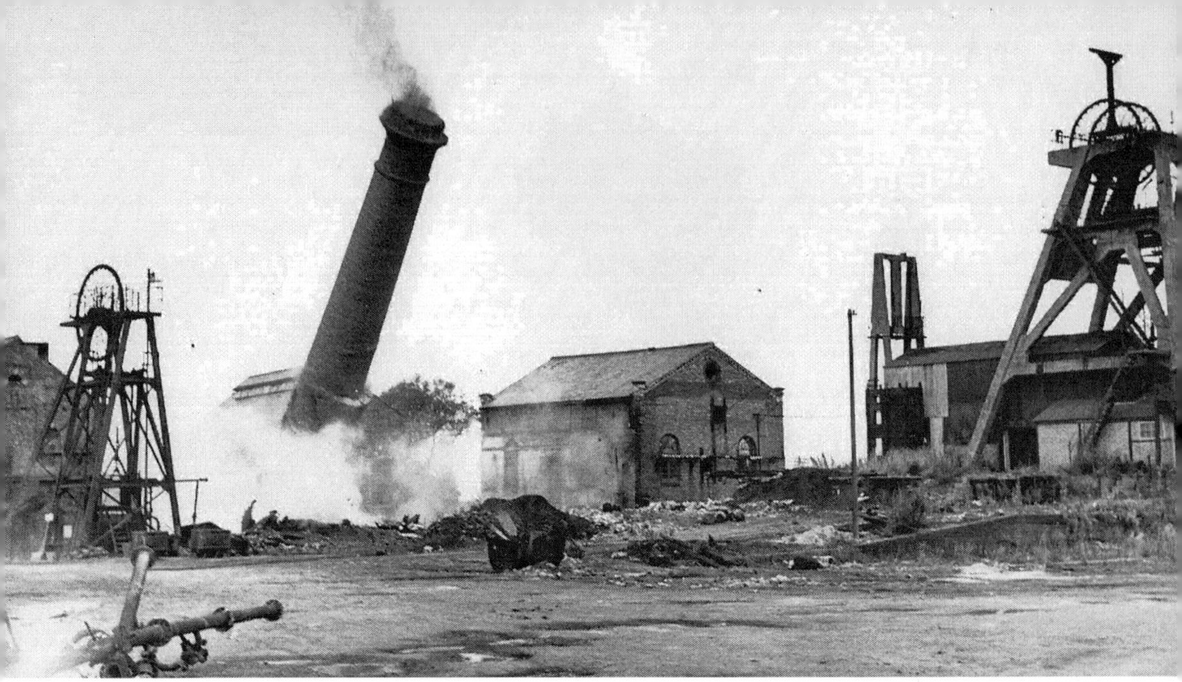

The demolition of Ansley Hall Colliery. A chimney is felled. (Reg Bull)

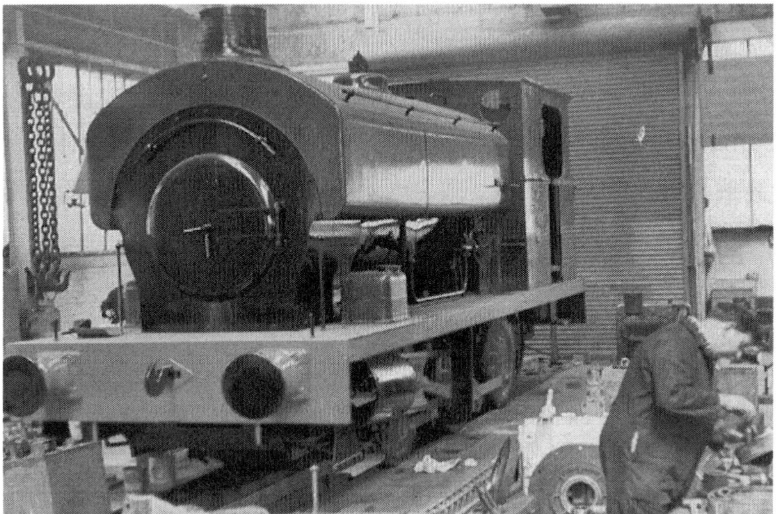

The NCB Ansley Hall area workshops were located adjacent to the colliery site and when built were connected by rail to the Stockingford branch. The workshops were opened in June 1959, but the rail link was short-lived since the branch closed with the colliery in November 1959. After the colliery was closed and the rail connection severed, repairs to steam locos were still carried out, as can be seen here when Haunchwood Colliery's big Avonside Engine No. 1883 of 1922 was turned out in immaculate condition after a long and leisurely overhaul. It arrived in March 1963 and was at Ansley until January 1967, when it was returned to Haunchwood for further use before it was moved to Arley Colliery on 13 September 1967. A year later, it was transferred to Newdigate Colliery on 27 September 1968. It moved again on 5 July 1969 to Baddesley Colliery. Then, on 26 July 1975, it was acquired for preservation by the Dinting Railway Centre, Glossop. This photo was taken on 7 May 1966. (Geoff Edmands)

THOMAS ASTLEY

Operated a blue brick and tile yard in the Black-a-Tree Gate Road (now Haunchwood Road) in the 1830s.

ATTLEBOROUGH ROAD BRICKYARD

This was extant in 1845 but may have been obliterated by the construction of the Coventry–Nuneaton railway line. It stood adjacent to Anker Street on what was then open land called 'Fudgy Nutts'. 'Fudgy' because of its soft clay ground, and 'Nutts' after the old local family of Nutt, who owned the land. The last Charles Nutt died around 1822. No brickyard operator is recorded, although it is likely to have been one of the Nuneaton brick-makers listed here. Soft clay subsequently caused difficulty for the railway company, as part of the brickyard lay under the locomotive shed yard and caused instability to the railway sidings laid over the top of it at a higher level. Frequent maintenance was needed.

ATTLEBOROUGH STONE QUARRIES

Known proprietors: Nathan Smart (listed as a stone mason, during the period 1828–41); Joseph Barnwell (listed as a stone mason, quarry owner by 1850); Nathaniel Smart (1791–1860); William & Joseph Seal (by 1863–71); Foxwell & Davies (quarry proprietors and brick-makers, by 1871 and 1884); David Parker (by 1888); Thomas Smith (builder, quarry owner, brick-maker, by 1892 and his firm lasted until the 1930s).

Thomas Smith (builder) built the Chase (now the Chase Hotel), Coton Arches, Newdigate Hotel, the former Midland Bank and other important Nuneaton buildings.

Rail connection: a siding was put in on the London & North Western Railway, Trent Valley line. In addition to stone traffic, the siding may have been used for local deliveries of coal for distribution in the Attleborough area. The main line connection was disused by 1900, although the signal box controlling it was in use until the 1960s as a fringe box for Nuneaton. There was also a short internal narrow-gauge tramway, possibly hand-worked or by horses, for removal of the stone from the working faces to the cutting area.

Canal: the Coventry canal was some distance away but was used for long-distance haulage of Attleborough stone, as deliveries of materials are known to have been made by canal from this source. Transport to and from the canal loading point was by horse and cart.

Attleborough freestone has been worked from medieval times. It is a form of sandstone, which was ideal for building purposes, being easy to cut and shape into regular-shaped blocks with little waste. It was durable and resistant to weathering. Trade directories described Attleborough stone as 'white, excellent freestone sent to various parts of the kingdom'. Many important local buildings were constructed from Attleborough stone. When these were demolished, the blocks were often reused for new building, either by being robbed out or sold on demolition. It was regarded as a high-quality material, and you would have seen buildings faced with Attleborough sandstone to the street, whereas the side and rear walls were in hard rubble diorite from the Hartshill quarries, or in brick from the local brickyards. Eventually, brick superseded Attleborough stone because it was much cheaper, more abundant, and easier to transport. There were dozens of brick kilns in the Nuneaton and Bedworth area. By the mid-nineteenth century, a brickyard was incorporated into the quarry. The stone was also used by monumental masons for headstones. The stone was actually light grey in colour.

Here is a list of a few identified buildings that used Attleborough stone in their construction, although there must be more still surviving in the local area:

- Horestone Grange, Nuneaton's medieval manor house erected in the ninth century but burnt down in around 1760 and the site robbed out. ('Horestone' means grey stone, and the manor was built with this.)
- St Nic(h)olas parish church, Nuneaton, and its perimeter wall
- Chilvers Coton Old School (now the Chilvers Coton Heritage Centre)
- All Saints church, Chilvers Coton
- The garden wall around the Bull Inn, Attleborough
- Coventry Cross, erected in the sixteenth century. (The cross was covered in gold leaf.)
- Royal Military Barracks and Ordnance Depot, Weedon, Northamptonshire
- Attleborough Hall
- Weddington Hall (thought to have been dismantled, the stone blocks numbered and shipped to the USA for reconstruction)
- Wolverton (Milton Keynes) parish church
- Witherley parish church
- Stretton-on-Dunsmore parish church
- Shilton parish church
- Rupert's Tower, Leicester
- Nuneaton Workhouse
- Hinckley parish church
- Atherstone parish church

(I suspect that church building and mansion house construction was a major source of business.)

The quarry used to extend a considerable distance from Lutterworth Road, adjacent to the railway, across to Bulkington Lane, with outlying pits worked sporadically, such as the 'Dumble Holes' near Teddy Kem's Heaven Farm, which was being worked in 1832 by a man named Everitt.

In 1871, a sale document lists the equipment at the quarry and brickyard:

> Engine house, boiler shed, several drying sheds, three kilns capable of producing 18,000 bricks each, three circular ovens, shedding, 25 horse-power horizontal steam engine, two boilers, brick making machines by Clayton & Shuttleworth, travelling crane, tramway, lifting crane.

The quarry was effectively worked out by the 1930s and has been partially infilled by domestic rubbish. It now lies under Paul's Land recreation ground. Part has been built over for housing development. The last piece of stone hewn from the quarry has been preserved in the front garden of a house in Quarry Lane.

Within the curtilage of the quarry was a windmill which ceased to grind corn in 1910.

CHARLES BALL

Builder and brick-maker, Attleborough stone quarry, 1871. Built Nuneaton Gasworks.

BEDWORTH BRICK TILE & TIMBER CO. LTD

Next to Bedworth station. Entered into an agreement for a siding with the LNWR dated 30 June 1900.

BEDWORTH COLLIERY

Owned by Sir Thomas Beaumont in 1615. Let to Sir William Turpin, 1615. John Bugge(s), Thomas and Anthony Robinson (1619). (These later mined on Newdigate land in the parish of Chilvers Coton.)

This colliery was owned by local landowner Edmund Anderson in 1570. In that year, it was offered to rent at 20 shillings per year.

After the leases were taken up, it is believed the Bedworth mine had almost a complete monopoly of the local trade for 40 miles around. The colliery made £300 per annum profit in 1602. It attained an output of 20,000 tons per annum before 1631. Newcomen engines were installed in 1725 and 1726.

BEDWORTH 'RAILWAY' COLLIERY & BRICKYARD

Extant in 1849, formally abandoned by December 1891, although it is thought to have been disused for sometime before then. It was adjacent to the present site of Bedworth railway station on the side where the goods yard was situated. The colliery fronted Bulkington Lane. There were seven shafts down to the Seven Feet, Slate and Lady coal. In 1896, Mr William Beasley Heap was running a brickyard from this address, which was probably trading as Bedworth Brick Tile & Timber Co. The brickyard appears to have become disused by 1902.

WILLIAM BINDLEY

In January 1825, Mr Bindley found clay in an area of Pool Bank Street, Nuneaton, between there and Wash Lane (now Queens Road) and commenced manufacturing bricks. Bindley, who was a builder, had originally bought the property for £40 in July 1824. He then secured additional areas of the Pool Bank Close. In addition, he built some houses and shops on this site, increasing the value of his security. He rented land from the Harpur family in 1828.

BLACK BANK PIT

Black Bank, Bedworth. Possibly a pumping shaft for Exhall Colliery nearby. Shafts close by – Speedwell, Exhibition, Balance Spinney and Old Arm Pits – were part of Hawkesbury Colliery. They were all worked out by the 1880s.

BONNER'S BRICKYARD

A small brickyard was operated in Wash Lane, Nuneaton (now 225–233 Queens Road). It was initially operated by Samuel Bonner. There was a small clay hole on the site. Following Samuel Bonner in 1867 was Arthur Hood, who worked the yard up until 1874. Hood lived at 7 Gas Street, Nuneaton. (Gas Street was later to become Queens Road at the town end.) In the 1870s, the brickyard was taken over by a railway contractor, William Moss of Stafford, who was engaged at the time in laying a third track down the Trent Valley main line for the LNWR. Moss had also rented the old Coventry–Nuneaton loco shed south of Nuneaton station, where he kept one or more contractors' locos for the duration of his contract. It is highly likely that Moss made bricks for the various culverts and bridges that he built on his contract at the yard in Wash Lane, and on termination of his contract he abandoned the workings. In 1883, this brickyard was offered to let. There were three fourteen-hole brick kilns, two large flame drying sheds, and two dwelling houses occupied by Whitehead and Boyer. The site was 256 feet long adjoining Wash Lane (now Queens Road). The area of the property was 2 acres. Brick-making ceased at this time, and the kilns were not used thereafter. The site was cleared. It was later built over as part of the housing and industrial usage in Queens Road. All traces were obliterated.

BOON & SON, GRANITE QUARRIES, HARTSHILL

Owned by: Richard Alkin; Joseph Boon; W. H. Bassett Green (trading as the Three Spires Granite Company) in the 1930s; Judkins Ltd from November 1955.

This quarry stood at the top of Tuttle Hill just below the windmill. It was served by a narrow-gauge tramway that went down

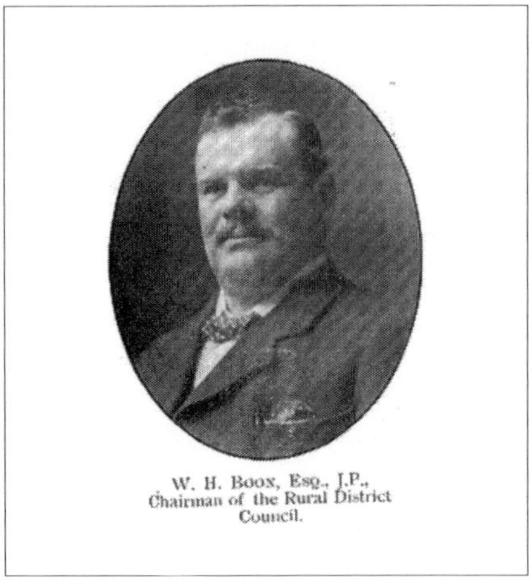

Above: W. H. Boon.

Right: Advertisement for William Boon & Sons.

Below: This plan shows the Nuneaton Corporation quarry on the left, and the more substantial Boon's establishment on the right.

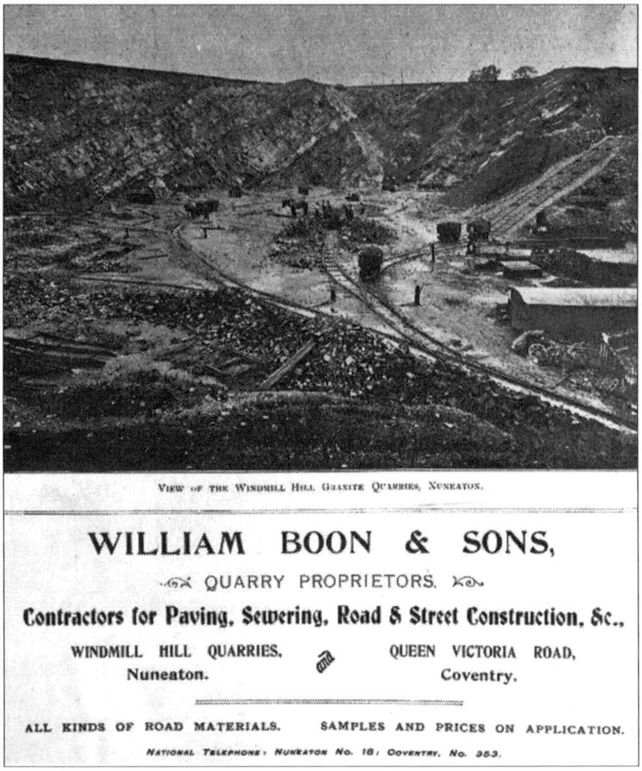

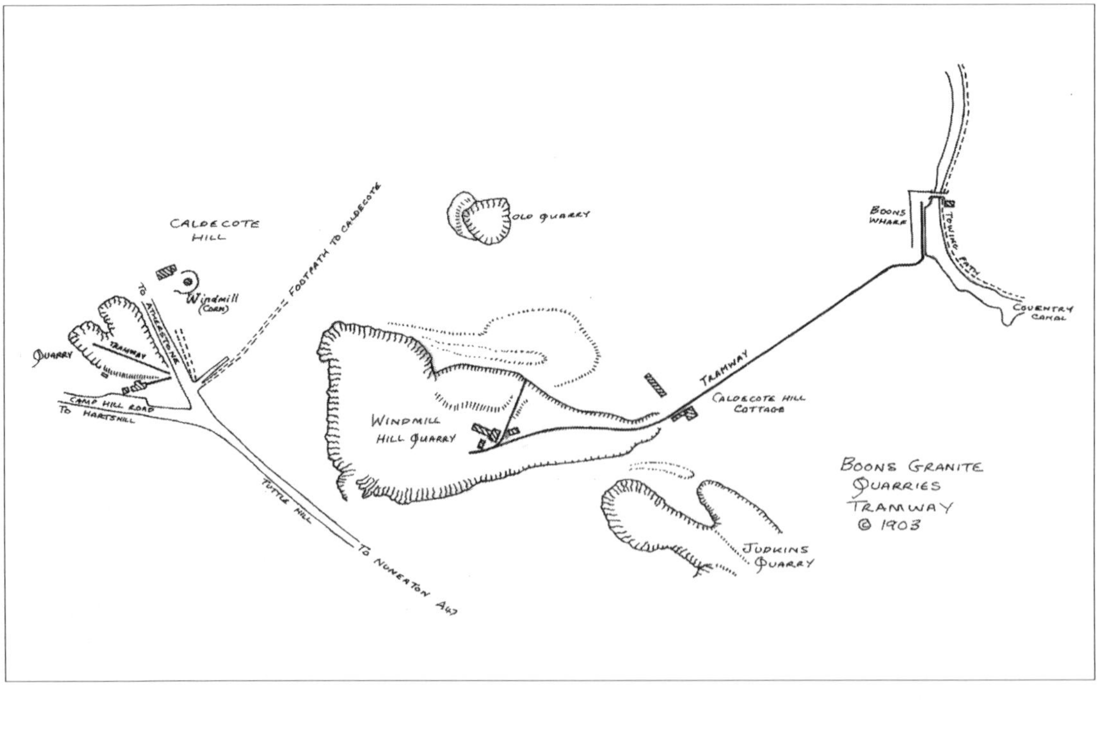

to a wharf on the Coventry Canal. At first, the railway was horse-worked until, in 1929, the company purchased a new Kerr Stuart DX-1-type, four-wheeled diesel named *Bassett Green*. During the ownership by W. H. Bassett-Green, the company were advertising that their stone had the highest crushing strain of any granite in the West Midlands. When Judkins took over, the narrow-gauge tramway was lifted between November 1955 and January 1956. *Bassett Green* was sold for scrap during 1957. Judkins extended their workings into Boon's Quarry, and the whole merged, obliterating Boon's operations.

BUNNEY BROTHERS

Extant in 1896. Brick-makers in Collycroft. Location of yard not known. Maybe working the brickyard of Collycroft colliery.

JOHN BUTTERTON

Extant 1542 as a tile-maker. Situated on a site near what is now Stanley Road, Nuneaton. Rent being paid: 13*s*. 4*d* per annum.

CHADWICK'S PIT, STOCKINGFORD (SEE NUNEATON COLLIERY)

Messrs Heath and Chadwick. There was an explosion at the pit in August 1868, during Heath and Chadwick's proprietorship.

CHARITY COLLIERY (OWNED BY THE NICHOLAS CHAMBERLAIN CHARITY, BEDWORTH)

Lessees:

- Peter Unger Williams (1775–1837)
- Caroline Williams (widow of P. U. Williams) (1780–1873) managed the colliery until 1858
- John McTaggart Williams (1810–1859) and Charles McTaggart Williams (1819–1858) trading as Williams Brothers (?–1861)
- George Barker & Sons, trading later as Bedworth Coal & Iron Co. Ltd (1878 by 1896)
- Stanley Brothers Ltd (1899–1924): closure date 13 September 1924; lease given up 20 April 1925.

There are known to have been ten separate shafts on the Charity estates. They were grouped into two Charity collieries on different sites. Charity No. 1 opened in 1776, in the appropriately named Coalpit Fields in Bedworth, and closed in 1830. Charity No. 2 was opened in 1830 at Collycroft, to help alleviate distress in the ribbon trade. The royalties helped to build Bedworth Almshouses and a school in Bedworth. The colliery employed thirty-eight men and eight boys in 1841.

Ironstone was also worked in balls and nodules, and blast furnaces were erected and commenced operations in November 1873. These comprised two large smelting furnaces 50 feet high and cased on the outside in wrought iron. Each was capable of producing 250 tons per week. In addition, there were two calcining kilns generated by four powerful engines driven by five powerful steam boilers. This was only a very short-lived affair; by 1879, the furnaces were out of use. The reason was that the type of ironstone produced was designed to produce high-grade iron. Wrought iron was being superseded in the 1870s by steel, so production of iron at Bedworth rapidly became uneconomic. The furnaces were demolished in 1894.

A tramway opened to the Coventry Canal in 1835. When the Coventry–Nuneaton railway line opened in 1850, a short branch line was put in which connected to the Charity Colliery site at Collycroft. Several steam locomotives were employed.

In 1874, three men were killed, and many injured in an accident at the colliery. In 1891, 360 men were employed. In 1896, while trading as Bedworth Coal & Iron Co. (executors of C. J. Schofield), 445 men were underground and 130 on the surface, working the rider, slate, and 7-foot seams.

Charity colliery employed 868 men in 1910. In 1920, it employed 799 men. By closure in 1924, 757 were employed.

Seams used were: Six Feet, Seven Feet (or Low Main), or Nine Feet and Two Yard.

On closure, 800 men were thrown out of work. Production of some of the seams worked continued, being accessed through Exhall and Newdigate collieries, and this led to the Charity operation ultimately becoming unviable.

CHARITY BANKS BRICKWORKS

After the closure of Charity Colliery, a brickworks was started on the same site. Later on, this expanded into the production of breeze blocks. Charity Banks brickworks was notable for the fact that the famous and much-loved comedian Sir Norman Wisdom was at one time the managing director. He was a leading shareholder from some time in the 1950s until the works was closed in 1972.

CHILVERS COTON POTTERY

Previously owned by a man named Coxon, this pottery (as it is referred to in the *Nuneaton Diary* of John Astley) was situated near the Cock & Bear bridge over the Coventry Canal. There was a wharf known as Wash Lane wharf. The site also had lime kilns located adjacent. The remnants of kilns stood on the wharf at least until the 1990s, but have now had modern housing built over their footprint. Clay was obtained from the Haunchwood and Whittleford area. Products made at the pottery are not known. It may be that the output was ornamental items rather than the usual bricks and tiles. The pottery was extant in 1837. Nothing further is known of it.

COAL PIT FIELDS COLLIERY

Two shafts in Coal Pit Field, Bedworth.

COCKSPUR COLLIERY, COLLYCROFT

It is not certain whether this was later just known as the Collycroft Colliery, but it was being operated in the 1730's, so the shaft was older than that listed at Collycroft Colliery.

COLLYCROFT COLLIERY

Collycroft Colliery was owned by William Wilson during a period in the 1830s until at least 1854. It was taken over by Edward Wilson, who was still operating it in 1876. It later traded as Howlette & Wilson Estates Colliery & Brickworks.

Believed to have been sunk in the 1830s and worked by a whimsy, the colliery was also referred to in one document as a 'scavenger pit'. First closed in 1857, it was reopened in 1873 due to the effects of the Franco-Prussian War, then closed in 1875 and abandoned in 1876. It reopened in 1893, with brickworks attached to it, although this might have worked independently previously. Before closing again in 1901, then reopening on a small scale in 1910 before formal abandonment on 28 March 1910, it worked the four-foot and seven-foot coal. The brickworks was closed in around 1902.

COVENTRY COLLIERY, KERESLEY

Operated by the Warwickshire Coal Company Ltd. (a subsidiary of the Coltness Iron Co. Ltd from 1924).

Although this colliery was on the borderline between the Borough of Nuneaton

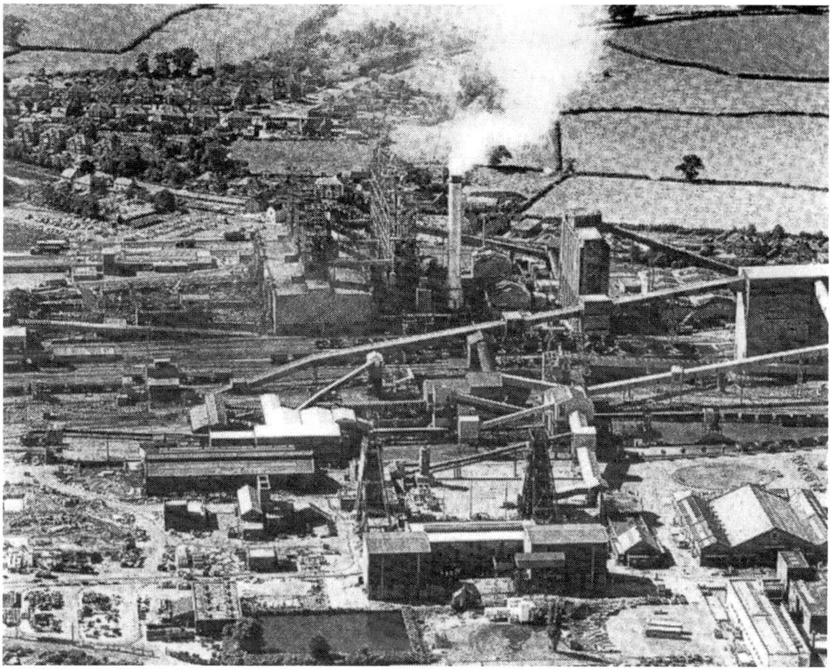

An aerial view of Coventry Colliery in the 1960s.

and Bedworth and the City of Coventry, now in the West Midlands, it is included because the largest part of the colliery itself lies within the Borough of Nuneaton and Bedworth, while the branch line down to Three Spires Junction is within the West Midlands conurbation area.

The company was first registered in 1901. The sinking of this colliery took four years, and the first coal was reached in 1917. Full production started in 1923. A branch line from the Coventry–Nuneaton line was commenced in 1911 (a siding agreement between the LNWR is dated 30 June 1911) and was two miles long. Extensive interchange sidings with the LNWR were put in at Three Spires Junction with a capacity for 1,600 wagons. The company owned 300 wagons and 700 on hire. There was also a land sale depot at Bedlam Lane. The canal wharf was at Longford Wharf on the Coventry Canal. On completion, the colliery was the largest and deepest in the whole of Warwickshire. It was built to access the Warwickshire Thick Coal (consisting of Slate Coal 6 feet thick, Ryder 6 feet thick, and the Two Yard), which is situated here at 720 yards deep. By 1923, output at the colliery was 20,000 tons per week from 4,500 acres of mineral rights owned or leased by the company. More than 200 tons of coal per hour could be conveyed, sorted and graded at the seams.

A company estate of houses was laid out with 350 new homes for 30 per cent of the 1,800 men employed, of which 350 were on the surface. Pit head baths were completed in 1934.

Coal produced:

- Two yard – large lumps – house coal
- Best screened cobbles – low ash content house coals
- Hand-picked bights – large lumps – similar usage as screened cobbles

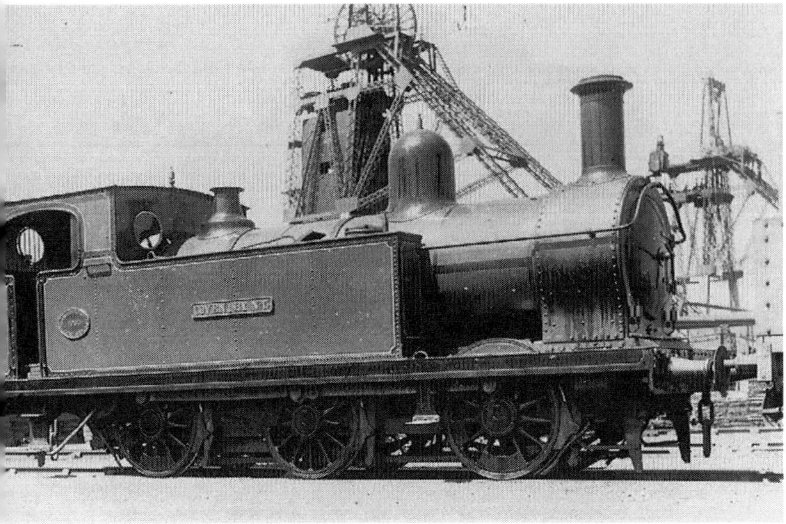

This was Coventry Colliery's star engine. *Coventry No. 5* was a 'Sharpie'. Sharp Stewart built many elegant locos and this is no exception. When new, it was built for the Barry Railway and was a very historic loco, well worthy of preservation. In this picture it looks in very good order. She was cut up on site by J & H. B. Jackson Ltd in March 1962.

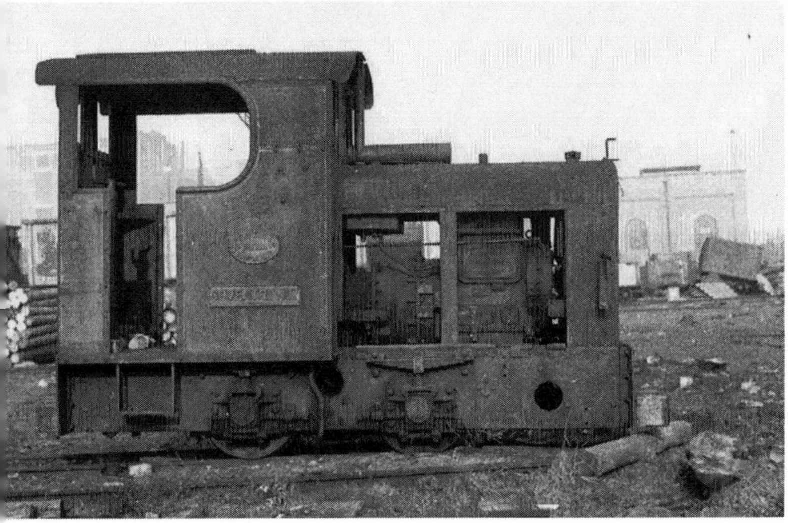

Keresley was a four-wheel diesel mechanical listed as *Coventry No. 7*. It was constructed by John Fowler (their No. 21915) in 1938, and was originally used for movement of stock coal. After 1947 and nationalisation it saw little use. Here it is lying derelict, with engine compartment doors removed. It was eventually sold to a local scrap merchant – G. Comley Ltd, in around late 1959 or early 1960 – and dismantled. (Industrial Railway Society/ K. J. Cooper collection)

- Large nuts – house coal
- Bakers nuts – fuel for Triplex and small grates
- Medium nuts – smaller version of the above for Triplex and small grates
- Best main deep – large house – kitchen cobbles – suitable for institutions
- Spire cobbles – long-lasting flame suitable for annealing furnaces
- Hand-picked steam-screened coal – suitable for steam raising, horticultural and agricultural purposes
- Double screen nuts – special – beans – nutty slack – ¾-inch slack – fine slack – steam-raising fuel with high evaporative power and low ash content

Coventry Colliery took delivery of two of these former GWR 0-6-0s. This was a very modern loco, which appears to be in excellent condition in this view of it at the colliery, possibly shortly after delivery from Andrew Barclay's in Kilmarnock, following overhaul in August 1962. It remained at the colliery until it was sold for scrap to J. Cashmore Ltd, Great Bridge, Staffs, on 23 October 1970. Its sister engine No. 1501 was preserved on the Severn Valley Railway.

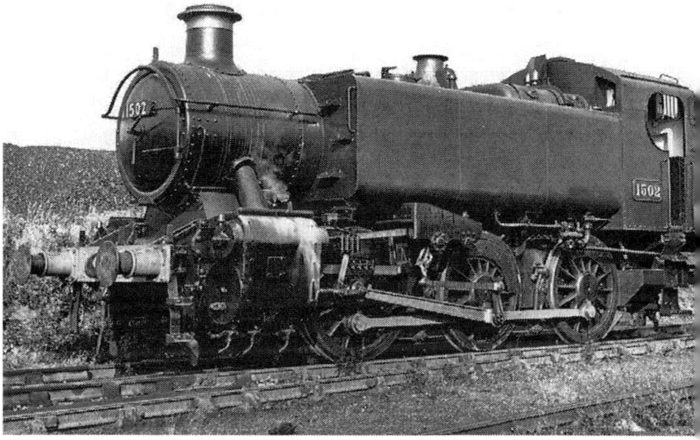

Coventry Colliery had coal sales agents covering London, Reading, Birmingham and Bristol. Most of the material produced, apart from local sales, was despatched to these areas.

A second land sale depot at the colliery itself was established. A Homefire Plant was erected in 1965.

The colliery closed on 16 October 1991 and was placed on care and maintenance in case of future reopening. The coal preparation plant and surface railways closed in 1993, and the site was sold to Coal Investment Ltd on 5 April 1994. The colliery had reopened in March 1994, but the financial results were poor and the colliery closed completely on 23 August 1996. The shafts were capped and filled.

The colliery branch line has been refurbished by Railtrack and is now used for trains to the Pro-Logis distribution site, which has been laid out on the former colliery. It is now possible to see main-line diesels accessing the former Coventry Colliery branch.

The history of the railways here is complex and it is recommended that reference is made to the Industrial Railway Society's *Industrial Railways and Locomotives of Warwickshire* handbook for fuller details.

WILLIAM COX, BRICKMAKER

Owned by Ruskin until 1879. Cox acquired the brickyard from a previous owner (Mr Ruskin) when the latter retired in May 1879. It is listed as a blue brick and tile yard in 1880, with Thomas Cox as manager.

CRAVEN COLLIERY

Lessees: Sir Thomas White's Charity, administered by Coventry Corporation, associated with, and took over from, Wyken Colliery, which was part of the same tenancy as Craven Colliery. In 1896, it was operated by W. S. & J. Harris, then by the Warwickshire Coal Co. Ltd.

The colliery was sunk on Serjeants Farm by 1837. In 1891, seventy men were employed. In 1896, there were fifty-nine men underground and twenty-eight on the surface.

Craven Colliery merged with Wyken Alexandra Colliery in January 1914, and closed at the same time as Wyken pit in October 1927 due to flooding. The connecting tramway was narrow-gauge. No locomotives are recorded, so the line may have been horse- or rope-worked.

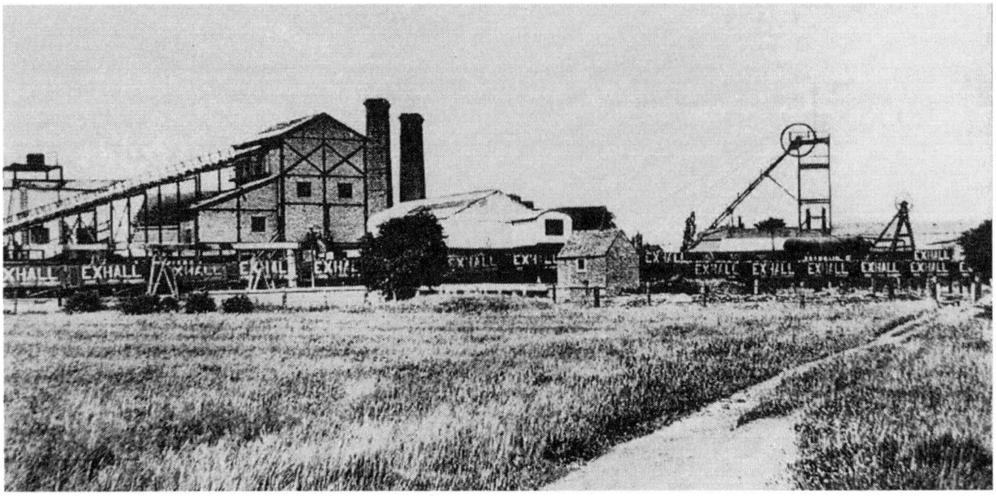

This picture postcard view of Exhall Colliery was taken between 1900 and 1910.

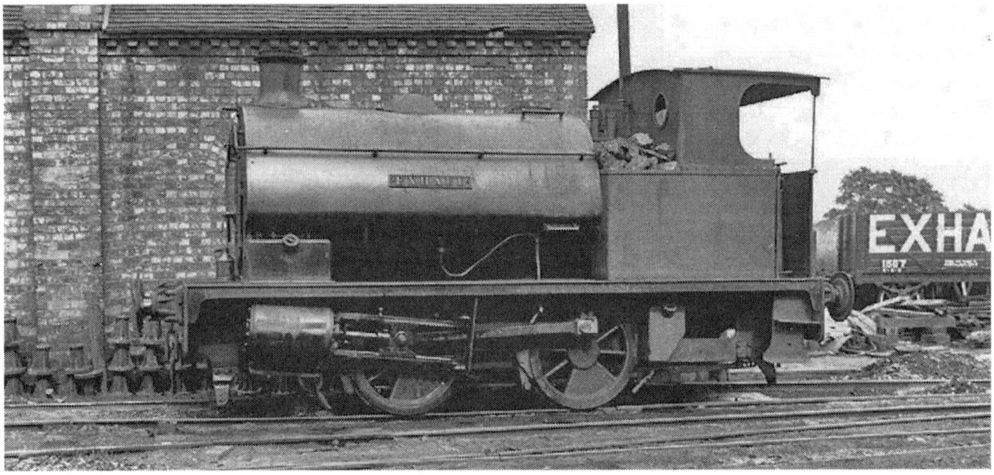

Exhall loco *Exhall No. 4* was one of two Bagnalls at Exhall and the first delivered new in 1912. It was Bagnalls works No. 1952 of 1912 and was sold for further service to the Leicester Colliery & Pipe Co. in 1938. (Industrial Railway Society/K. J. Cooper collection)

EXHALL

This was Exhall Coal Mining Co. Ltd (Edward Wilson) from 1857 until around 1896, then it was Exhall Colliery Co. Ltd until 21 December 1906. Between 1906 and 1939, it was Exhall Colliery & Brickworks Ltd, then Hawkesbury & Exhall Collieries Ltd.

Mining has been carried out at Exhall for several hundred years, and we know that in the year 1774 there were eight individual shafts. A new shaft was being sunk in 1858, but was abandoned in 1860. Then two more were started and coal was being raised in

1861. Sinking seems to have been completed in 1864. Temporary closure occurred in 1869. Flooding may have been the problem. Coal extraction seems to have commenced by 1882. There was a temporary closure due to flooding in 1887.

Three brick works were also associated with the site. No. 1 yard closed in May 1916. Exhall had an average weekly output in 1929 of 70,000 bricks. 3,674,000 bricks were made in 1930. The brickworks losses started to accumulate in the 1930s and this led to closure of their brick yard interests on 11 September 1939. Three men stayed on at the brick yard until 1941 to dispose of 1 million bricks in stock. The clayhole was near Speedwell Farm.

In 1891, 270 men were employed in the colliery. In 1896, there were 351 men underground and 125 surface workers. In 1914, there were 910 men underground and 200 above. The colliery mined the Ryder, Ell and Slate Coals. Underground workings were extensive – some headings were two miles from the pit bottom.

The colliery closed on 3 April 1938 due to increasing losses. Nine hundred men were laid off. It reopened in June 1939, but was closed in 1943 due to a fire. It subsequently reopened, employing 200 men before final closure in February 1949, again due to fire. In 1947, 49,506 tons of coal was raised. In 1948, 38,592 tons of coal was raised. The pithead buildings stood until the 1960s, but were later demolished as the site disappeared under the Bayton Road Industrial Estate.

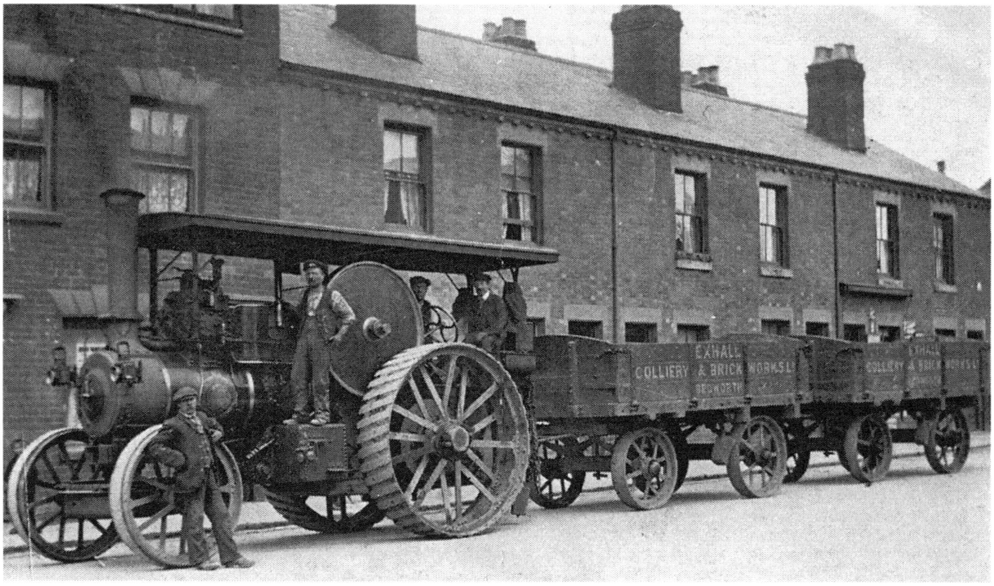

A Burrell steam traction engine, hauling two road wagons delivering bricks to local building sites in the Coventry area, had a chequered career. The engine was works No. 2081, delivered new on 26 April 1898 to Troward Clark of Eastry, Kent. By 1910, it had switched ownership to A. V. Davies of Avenbury, Hereford, being sold through Wingham Engineering of Wingham in Kent. After 1910, it was sold to Exhall Colliery, and they then sold it to Pat Collins, the amusement company at Bloxwich, Staffordshire, by 1918. In 1920, it was sold by them to J. Smith, a dealer, and was subsequently scrapped at an unknown date. This photo pins it down to the period around the First World War.

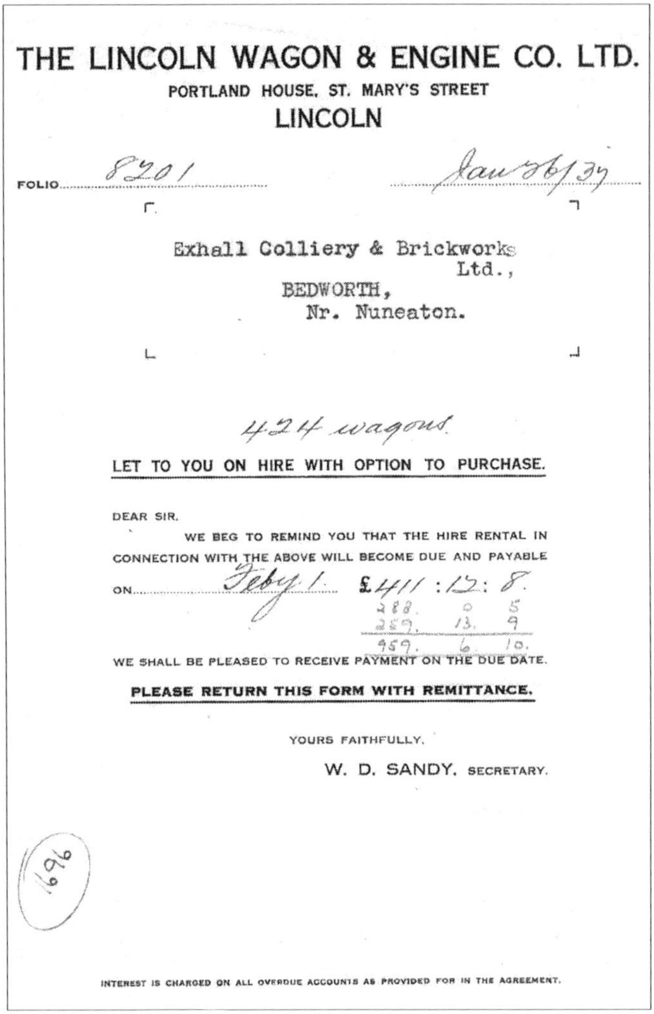
THE LINCOLN WAGON & ENGINE CO. LTD.
PORTLAND HOUSE, ST. MARY'S STREET
LINCOLN

FOLIO 8201 Jan 26/37

Exhall Colliery & Brickworks Ltd.,
BEDWORTH,
Nr. Nuneaton.

424 wagons

LET TO YOU ON HIRE WITH OPTION TO PURCHASE.

DEAR SIR.
WE BEG TO REMIND YOU THAT THE HIRE RENTAL IN CONNECTION WITH THE ABOVE WILL BECOME DUE AND PAYABLE ON Feby 1. £411 : 12 : 8.
288. 0 5
259. 13. 9
959. 6 10.

WE SHALL BE PLEASED TO RECEIVE PAYMENT ON THE DUE DATE.

PLEASE RETURN THIS FORM WITH REMITTANCE.

YOURS FAITHFULLY.
W. D. SANDY. SECRETARY.

1696

INTEREST IS CHARGED ON ALL OVERDUE ACCOUNTS AS PROVIDED FOR IN THE AGREEMENT.

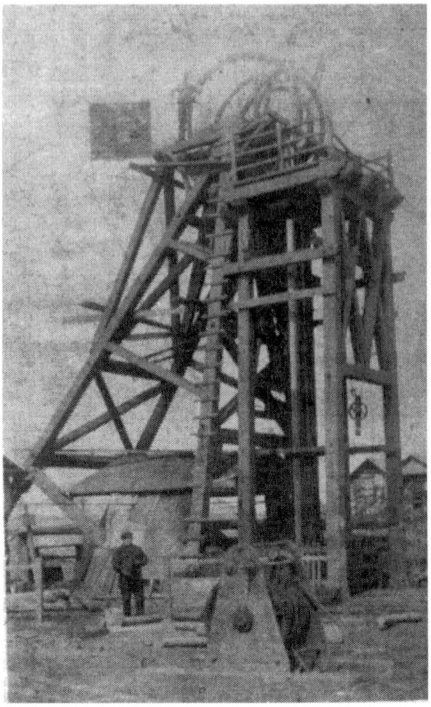

Left: Invoice for Exhall Colliery.

Above: This is believed to have been an old shaft at Exhall, as it came from the collection of a former employee now passed on, but we do not know for sure. (*Heartland Evening News*)

ROBERT GARNER

Garnett was a brick-maker who valued the inventory of his fellow brick-maker in Nuneaton, John Alexander, in 1680; that is how we know about him, but details of his operation and the dates that he was in business are unknown.

HENRY GREENE

Greene was a brick-maker (brick-making was listed as his profession in his will of 1637) but the location of his brickyard is not known.

GRIFF COLLIERY (LATER GRIFF COLLIERIES LTD)

The Newdigate family leased the mineral rights on their estate to coal-mining entrepreneurs, or worked the coal mines themselves. Lessees included:

- Matthew Collins, John Potter et al. (1622)
- George Sparrow & Richard Stonier Parrott (1715, 1720)
- Newdigate & Sparrow (1720, 1722)
- Richard Newdigate (1722, 1729)

A list of known shafts, pits, and collieries and their operating dates:

- Griff Delphs (1604)
- South Griff (closed by 1800)
- Griff Lowsey Meadows (1787 – *c.* 1800)
- Griff Engine Pit
- Deep Foundation Gin
- Barrel Gin
- Wheel Pit
- Horse Gin
- Whimsey
- Sir Richard Newdigate's Bassett Engine Pit
- Three Gins
- Mr Wilkins Gin
- Griff Barbara (1780–1850)
- Charles ? (extant 1880s)
- Caroline (1832–50, converted to a pumping station)
- Griff Slate Colliery was being worked in 1841
- Griff No. 1
- Griff No. 3
- Griff No. 4 (1851-22.7.1960)
- Griff No. 5 (Metz's pit, 1870–1900, named after the Battle of Metz)
- Griff New Winnings or Clara Colliery (1891 – 27 May 1955)

Mining had been carried on in the Griff area of Chilvers Coton for over three hundred years. Records in the Newdigate papers indicate coal pits being sunk in 1603/04. There were as many as eight separate shafts being worked at one time, but this was reduced to two by 1718. Later in the eighteenth century the Griff Colliery's trade seems to have fallen away before reviving after the Coventry Canal was completed.

Griff (New Winnings) later known as Griff Clara Colliery commenced sinking in 1891. A new colliery village was laid out, consisting of seventy cottages built at an average cost of £100 each. Seven blocks were in three-up, three-down terraces, the remainder in two-up, two-down. Bermuda village was erected in twelve weeks. There was a mission hall, a working men's club and women's institute. Edward Melly suggested it was named Bermuda village in deference to Lt-Gen. Sir Edward Newdigate-Newdegate, who was Governor of Bermuda from October 1888 to June 1892. In 1896, there were 494 underground miners and 166 on the surface. Clara worked the Two Yard coal.

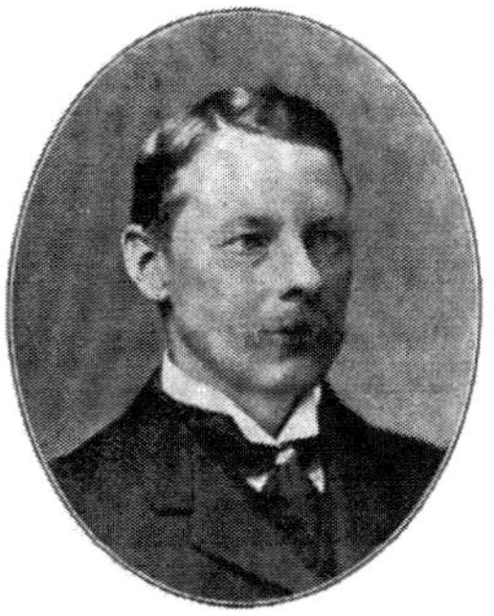
E. F. MELLY, ESQ., J.P., Member of District Council.

E. F. Melly.

Thomas Newcomen agreed to act as a drainage engineer to Griff Collieries in 1714. A Newcomen steam engine was erected in 1714. Prior to the erection of this engine, fifty horses were used to pump water out of the mines. Two more engines were erected in 1719 and 1725 respectively. It is thought that the original 1714 was dismantled and moved to Hawkesbury pumping station.

Underground the coal was moved in small tubs constructed on sleds that were pulled by 'drawers'. These were usually children. In 1777, an unusual device for moving the tubs underground occurred when a waterway was built connecting outlying areas of the pit to the bottom of the shaft. Later on,

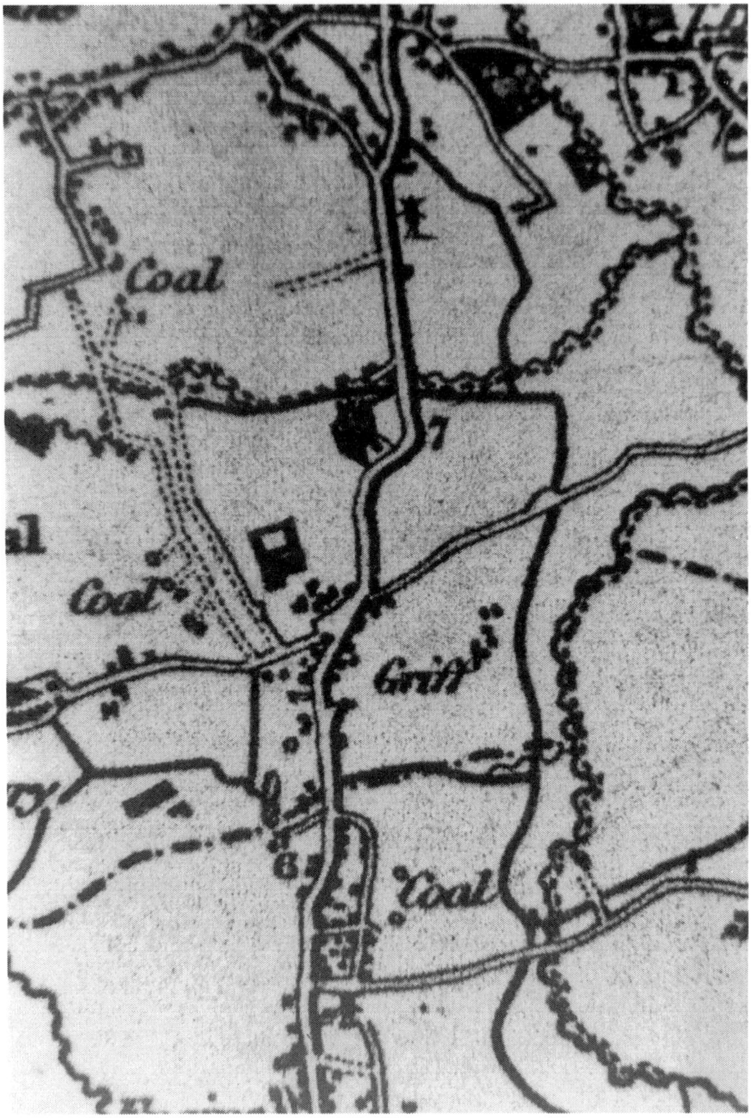

A late eighteenth- or early nineteenth-century map of Griff. The road in the centre is the current road that winds through the hamlet of Griff from Nuneaton to Bedworth. The map illustrates numerous outlying shafts along the outcropping coal measures. Coal could be found just a few feet below the surface in places here. The road off to the right at the bottom of the map is Marston Lane, Bedworth, and the cluster of houses above is the hamlet of Collycroft. The two shafts shown just to the right of these houses may be the Collycroft collieries listed elsewhere. At the top of the plan is the village centre of Chilvers Coton. (Fred Phillips collection)

underground tramways were adopted. On the surface, 'rail-roads' or tramways were adopted to move coal from the shafts to the canal wharves, or to the land sale. These tramways were reported to be of 3 ft 4½ in gauge. The iron rails and wheel sets, as well as iron pipes, cylinder castings, boilers and other castings, are known to have been provided by the Coalbrookdale Company.

Three small canal boatways were built to convey from outlying areas of the coalfield between 1700 and 1711. These were connected to the Coventry Communication Canal in 1768. In 1790, the canal was extended from Atherstone to Fradley Junction.

The Griff Colliery spur was opened in 1850; this connected several horse-drawn tramways to the Nuneaton–Coventry branch of the

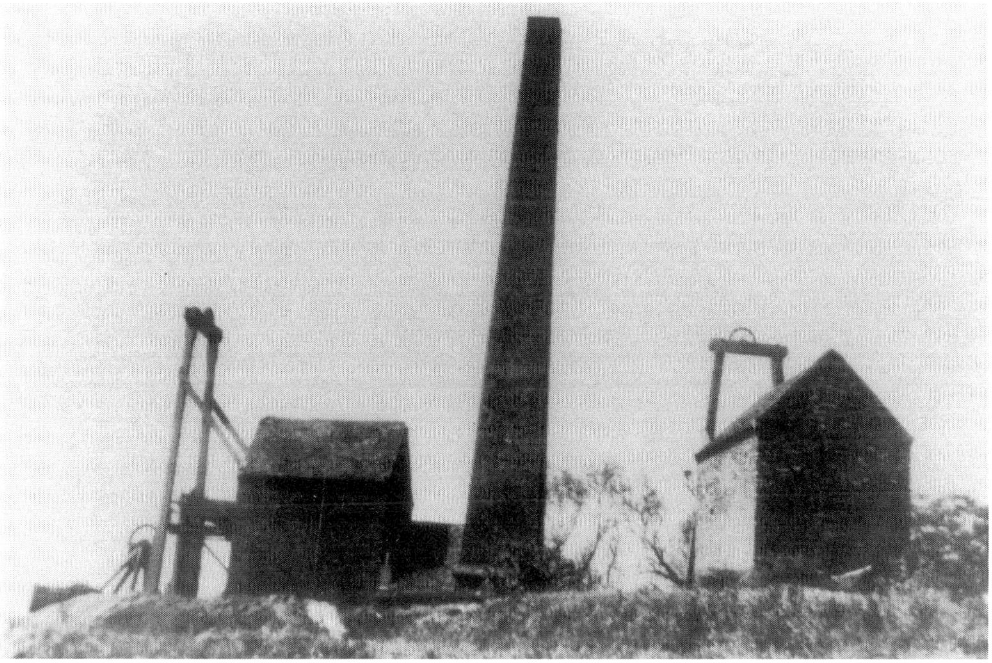

Caroline shaft, seen here, was converted into a pumping shaft. It was originally sunk for coal winding in 1851.

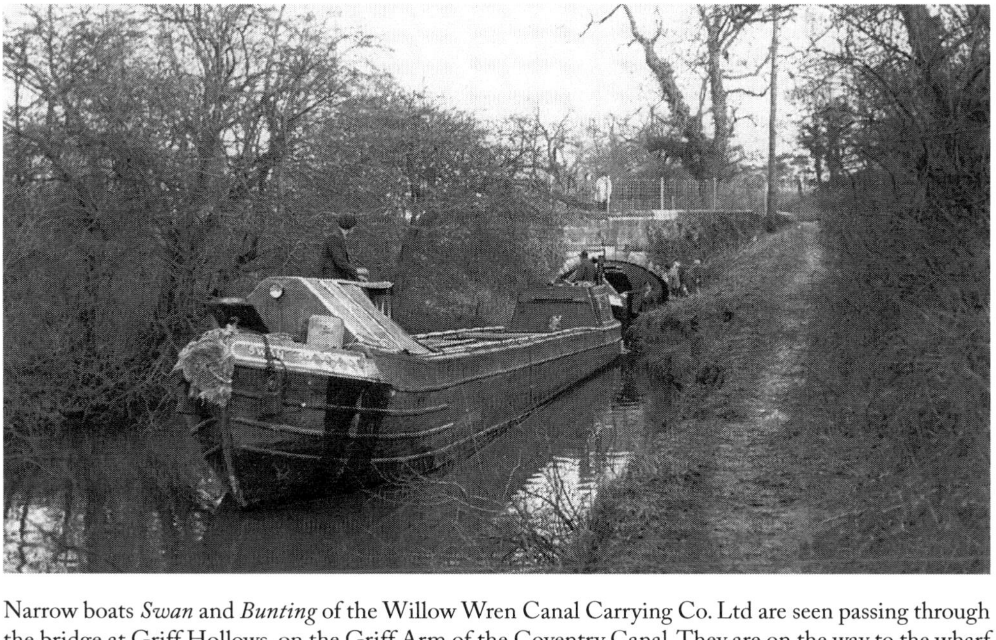

Narrow boats *Swan* and *Bunting* of the Willow Wren Canal Carrying Co. Ltd are seen passing through the bridge at Griff Hollows, on the Griff Arm of the Coventry Canal. They are on the way to the wharf at Bermuda village to fill up with coal. The date is 21 February 1960. This picturesque scene has been erased from the map, as the canal has been filled in. (Geoff Edmands)

LNWR. The spur was extended through to the Stanley Brothers yard at Stockingford and officially opened on 22 June 1881. The Midland Railway had running powers over the LNWR Griff branch from 11 September 1889.

In 1891, 450 men were employed. In 1910, there were 524 men working. In 1950, there were 550 men underground and 130 on the surface. In 1960, 527 were employed underground and 153 on the surface. In 1957, 187,300 tons of coal were produced.

Griff Engine Pit in 1896 employed 232 men underground and 127 on the surface. It worked the Slate coal.

Griff No. 5 Colliery was sunk in 1870. In 1896, it employed 125 men underground and 49 on the surface. It worked the Slate and Seven Feet coal, and in 1900 employed 65 men.

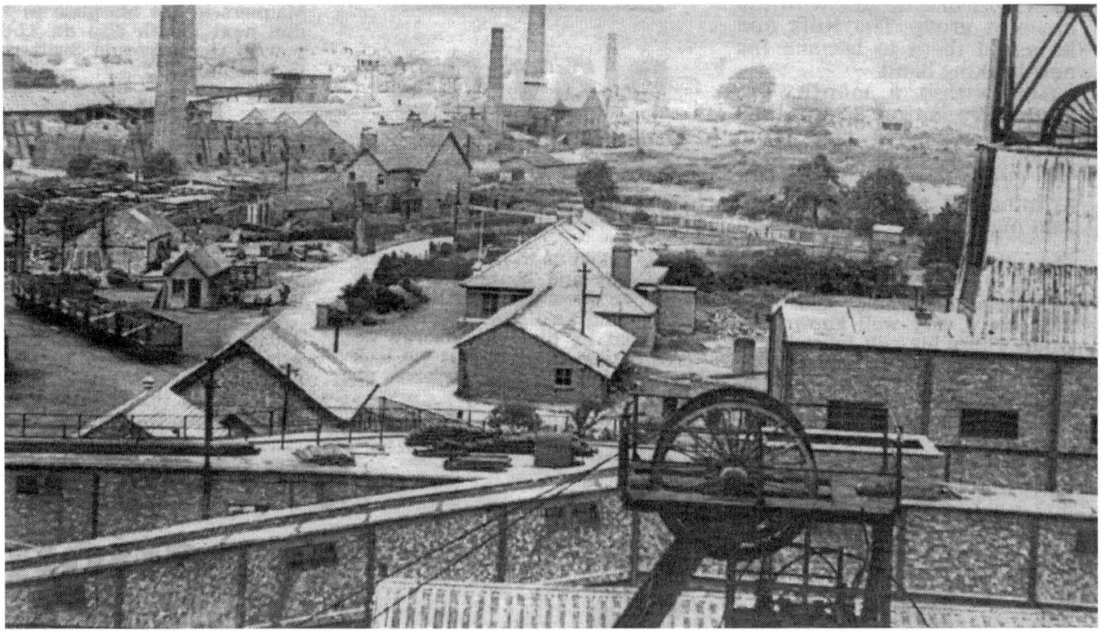

Above: Griff No. 4 pithead post-nationalisation. In the centre of the picture is a house fronting Heath End Road, which remains today and serves to pinpoint the location. Beyond the roof of the house to the left can be seen the serrated gable ends of Haunchwood No. 2 yard with chimney stacks. Griff No. 5 (Metz) colliery was located on the wasteland to the right. In the centre foreground of the picture is the single-storey office block.

Opposite above: Griff Charles shaft, *c.* 1884. The loco is a little four-wheel Black Hawthorn, *Swallow*. Works No. 174 of 1873. *Swallow* had an unfortunate destiny. It was sold to Neville Druce & Co. Ltd, Llanelly, South Wales, via Williams, Cornforth & Co., Kidsgrove, Staffs. Under new ownership in Wales, its boiler exploded, killing the footplate crew. In this view it is surrounded by the old-fashioned dumb-buffered wagons. Note too that the coal blocks are laid in the wagons neatly stacked as in a dry stone wall. The loaders in their heavy aprons had to ensure the lumps of coal, which were sold by size as well as quality, arrived at the customer in the size as purchased. Being reduced to slack by rough shunting and the sharp buffeting of these crude wagons would reduce the value of the load, and the deficit would be debited from the loader's modest pay. Packing the coal in these wagons to ensure the product arrived safely needed great skill by these hard-working men.

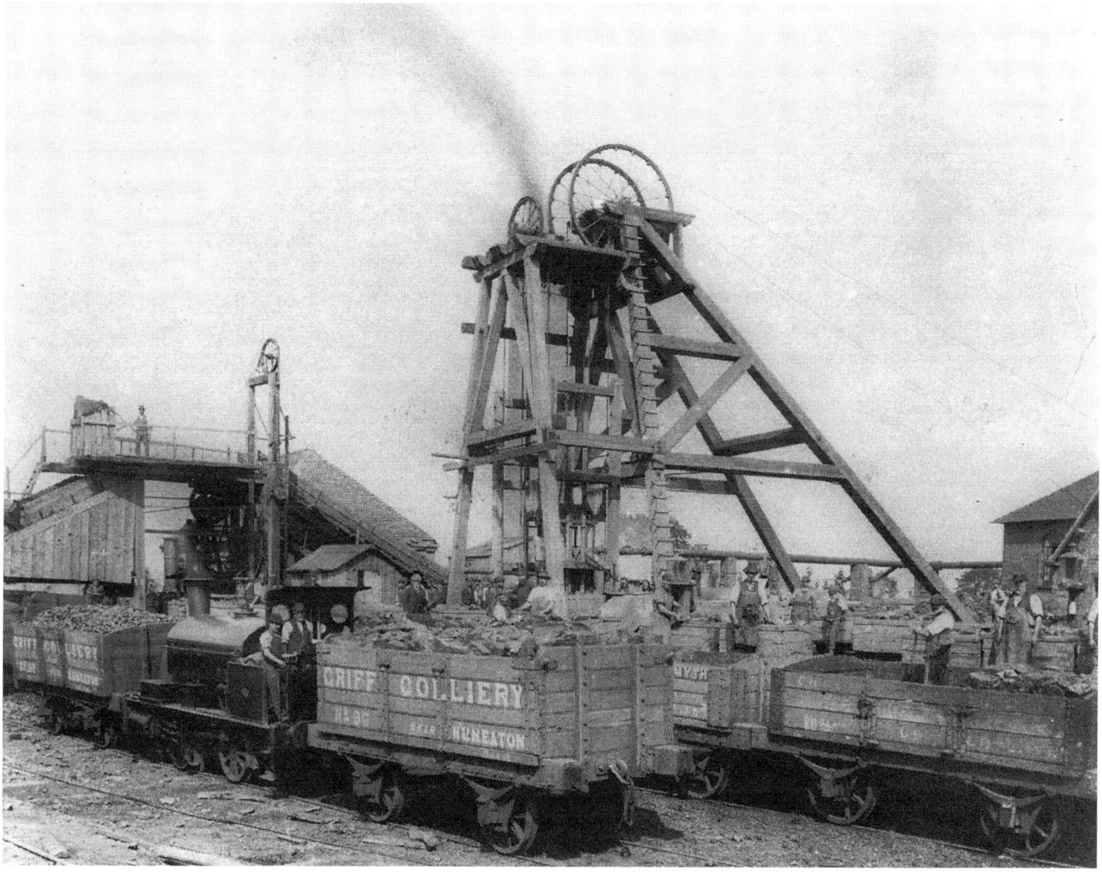

Right: Sinking Griff Clara Colliery in 1891. The full sinking operations took three years.

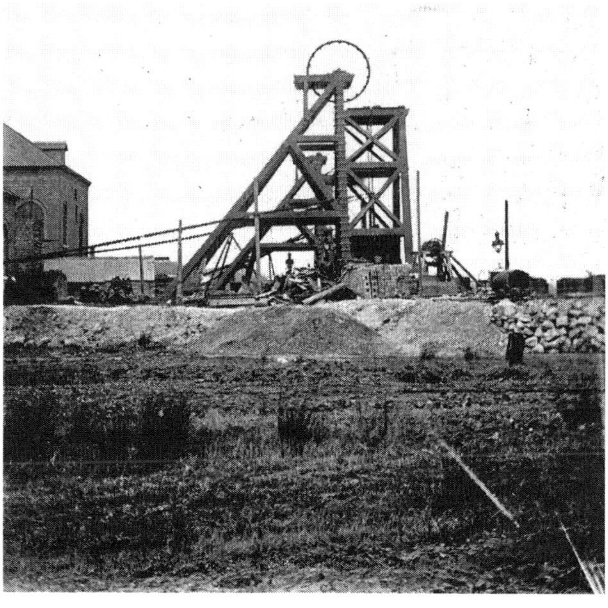

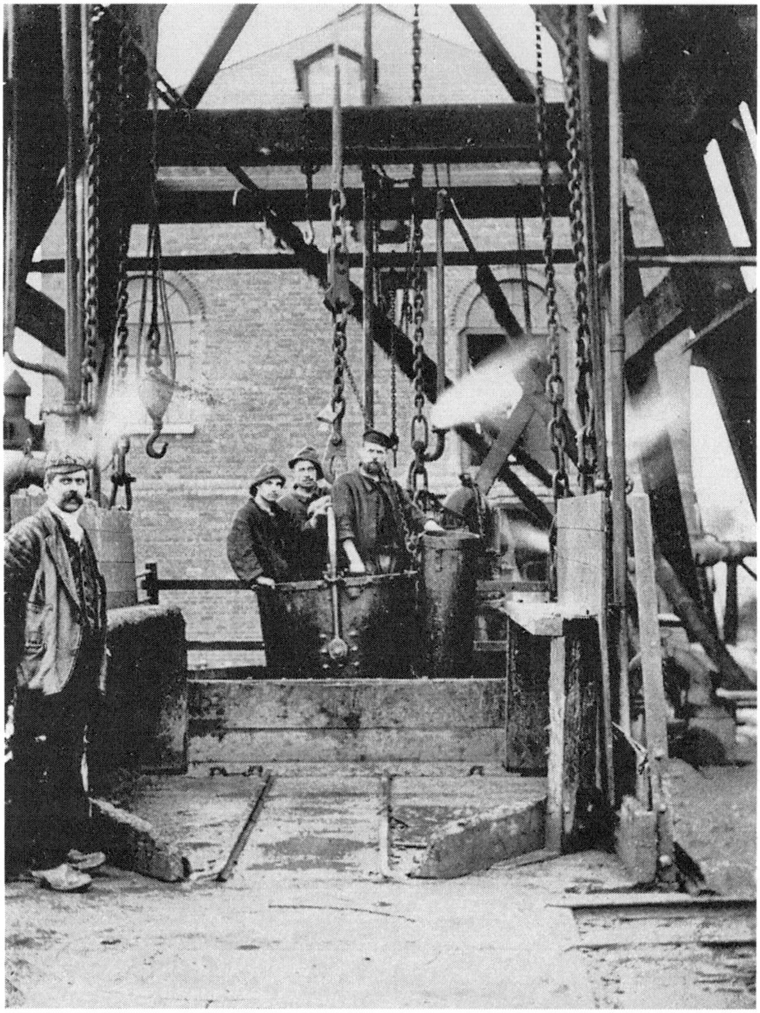

In this picture of the Griff Clara sinking under way, men are posing in their steel bucket prior to its being lowered into the shaft. No hard hats for these sinkers. Slouch hats seem to be favoured.

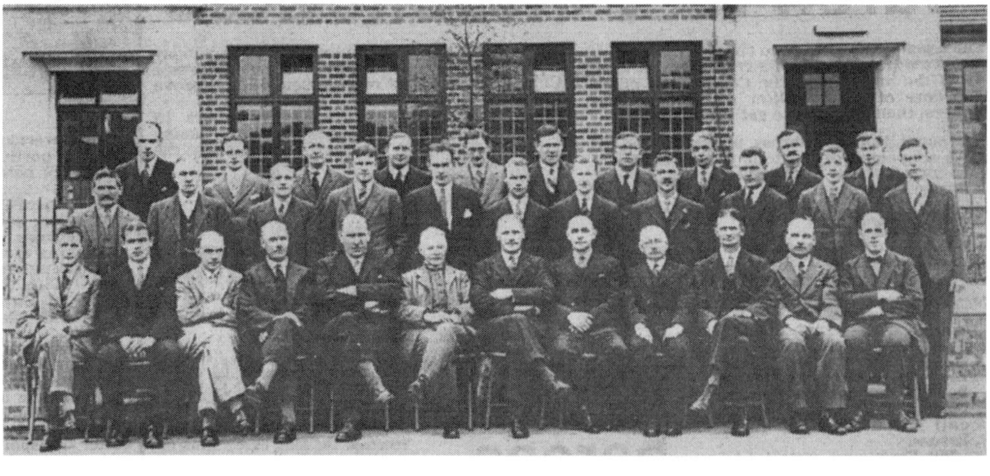

Opposite below: Senior staff at Griff Collieries outside the offices at Griff No. 4 colliery in 1931 to mark the opening of the new office block in the background. Back row from left to right: H. Jackson, J. Swann, T. Barson, J. Clarke, H. White, H. Paul, T. Knott, F. Goff, G. Moreton, and D. Clay. Middle row: H. Gunn, T. Waller, J. Clifford, E. Cleaver, G. Mawby, G. Dalby, F. Sharrod, B. Hutchinson, R. Thistlethwaite, D. Wright, and K. Daulman. Front row: V. Shrigley, F. Harrison, G. Neath, A. Mawby, F. Povey-Harper, E. Melly (managing director), H. Cleaver (secretary), C. Darlison, R. Clery, F. Orme, H. Lawrence and G. Watson. (From a photograph in the collection of Reg Bull)

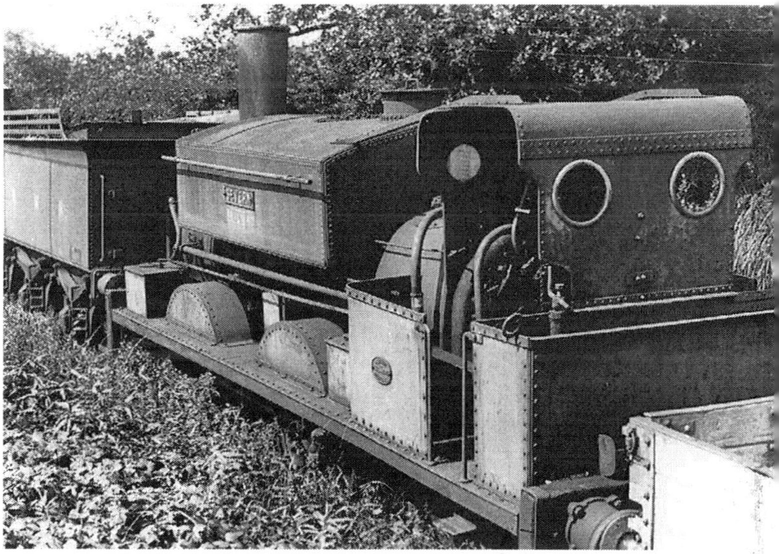

Among the fossilised relics gently rotting away among the weeds at Kinnersley on the Shropshire & Montgomeryshire Railway in the 1920s could be found this old stager – formerly a Griff Collieries loco. It was originally believed to have started life on the London & Birmingham Railway in the early 1840s, one of Bury, Curtis & Kennedy's pioneer locos for that line, although its origins are still in doubt. One school of thought seems to suggest it came from the St Helens Railway and was used by a contractor, Barnes & Beckett, on the Ashby & Nuneaton line contract, being put up for sale when that line was completed in 1873. The *Engineer* and the *Nuneaton Chronicle* in 1875 record that a former St Helens loco was sold by the LNWR to Barnes & Beckett in December 1871 for £450. It was originally St Helens Railway No. 6 *John Smith*, rebuilt in 1855 and said to have been constructed by Edward Bury in the 1830s or 1840s as an 0-4-0 tender engine. At any rate, it seems to have turned up at Griff in the mid 1870s and worked there for another thirty-odd years. Its real claim to fame came about when it was purchased for £350, then rebuilt and reconditioned by Bagnalls in Stafford, by the light railway engineer and entrepreneur Colonel H. F. Stephens, to work his newly reopened Shropshire & Montgomeryshire. Probably at least third-hand by the time Col. Stephens bought it May 1911, it worked traffic on that line before being laid on one side in a thoroughly worn-out condition. Like many of its fellow Col. Stephens locos, it was not cut up immediately, but simply rotted away. Parts disappeared and others were scattered until, sometime in the 1930s, the last pieces were hidden by the boscage. Perhaps components of the engine may still be found by someone who knows where to look among the undergrowth. (H. C. Casserley)

Next page, top: In 1948, a former War Department loco was obtained: a Hudswell Clark loco, No. 1102 of 1915, originally built for the Port of London Authority. It was requisitioned by the War Department and became their No. WD70211. It was well thought of at Griff, and a regular at Clara Colliery. However, by 1954 it was in poor condition and needed a thorough overhaul. By the summer of 1955, a quantity of spares for it had been purchased and dismantling began in September 1956. But by then Clara pit was shut and the impetus to repair it lost. The remains stood in the colliery yard in a derelict state before it was cut up for scrap in September 1960 by John Cashmore Ltd of Great Bridge, Staffs. Throughout its time at Griff, it retained its WD identity. The spark arrestor was fitted sometime after June 1952. (M. J. Lee)

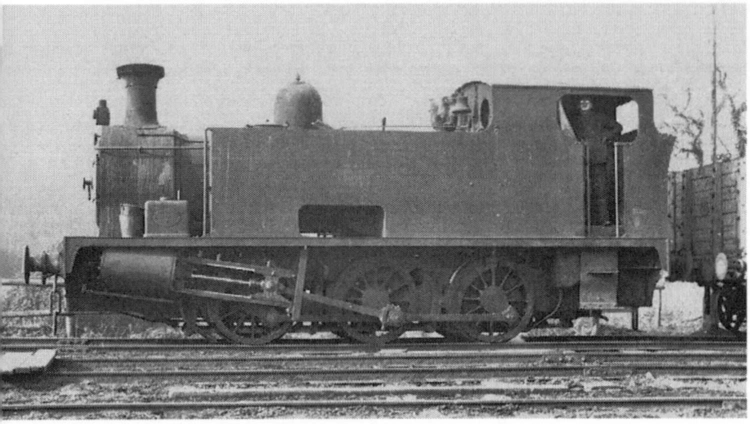

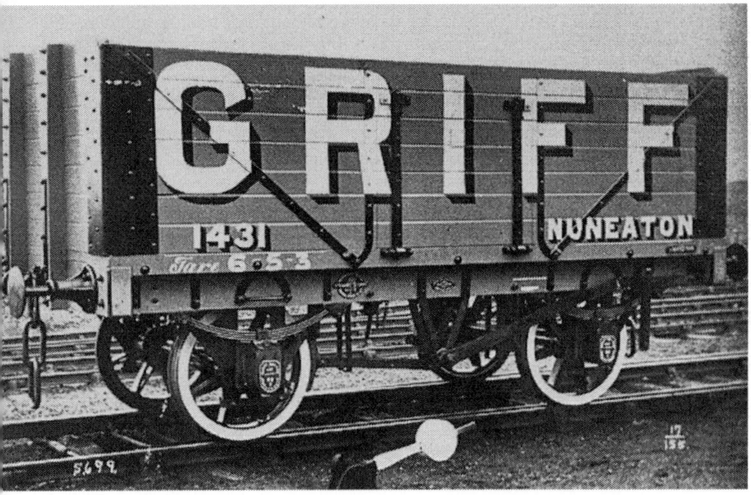

In common with all the local collieries extant in the early 1900s, Griff maintained a fleet of well over a thousand of these seven-plank coal wagons. There was a wagon repair shop at the colliery to maintain the fleet in good order. Major repairs could be carried out, and it is thought that complete wagons were assembled using bought-in metal parts. These distinctive wagons could be seen in trains and sidings all over the country. (HMRS)

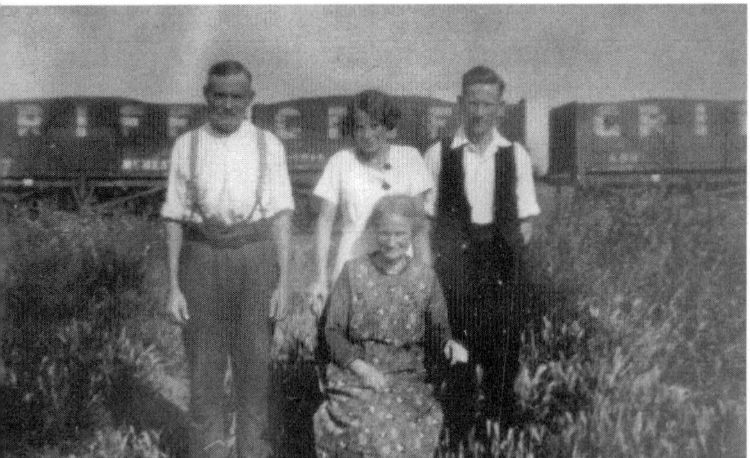

It is certain that someone will recognise this Bermuda family, with a fine rake of loaded GRIFF wagons, ready for despatch, as a backdrop. (A. Tunnicliffe)

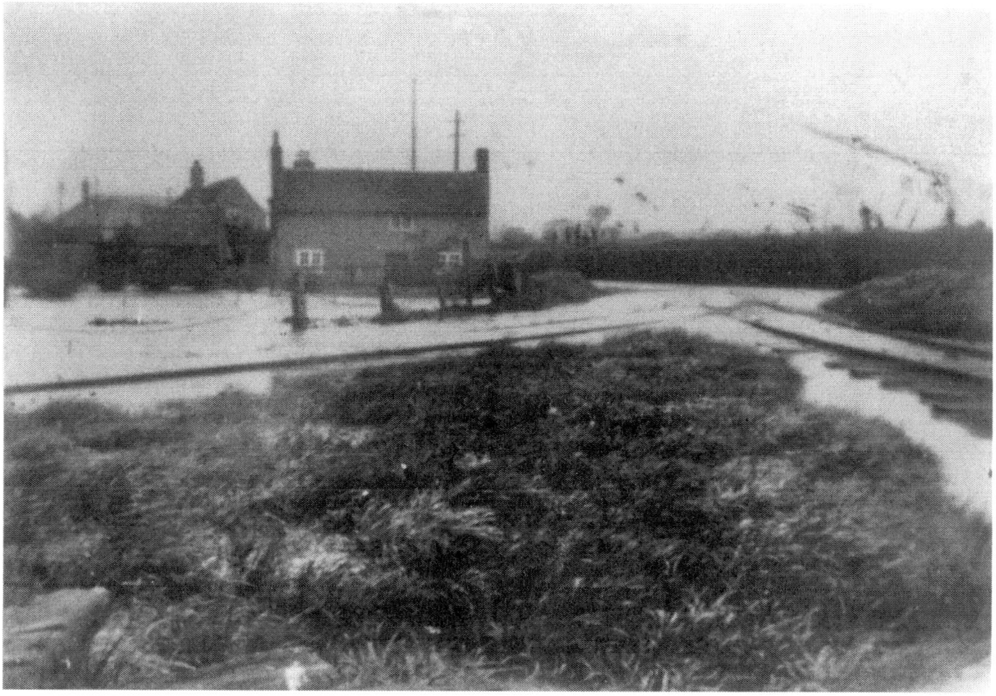

This was the scene adjacent to the Griff canal basin, with a certain amount of flooding to be seen. Railway tracks head off over the field to Clara on the left. (A. Tunnicliffe)

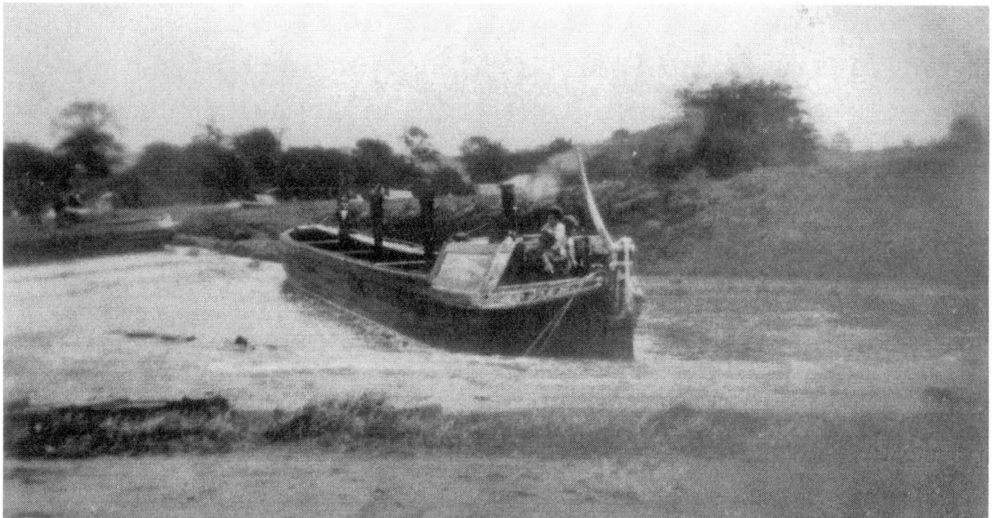

A typical scene at the Griff canal basin with an empty barge turning, ready for loading. (A. Tunnicliffe)

GRIFF BRICK, TILE AND LIMEWORKS

Thomas Kydd (1828–1892) was the proprietor. This facility was located at the canal wharf at the head of the Griff Arm of the Coventry Canal. It seems probable that the facilities here were worked before Mr Kydd purchased the business.

GRIFF GRANITE QUARRY COMPANY LTD

Based at Griff, Nuneaton. Known as the Griff Granite Quarry Company until 1908, and William Griffiths & Co. Ltd thereafter. The quarry was worked on a small scale from about 1850. There was a wharf on the Griff Arm of the Coventry Canal. Stone extraction gradually extended and a tramway on the 2 ft 6 in gauge was installed to connect to outlying excavations. In 1912, a private siding was installed that connected the quarry to the LNWR Coventry–Nuneaton line. The workings eventually connected to an extensive quarry in Gipsy Lane at Griff. In 1963, the narrow-gauge tramway was abandoned, and the single Motor Rail diesel mechanical loco was sold for scrap. The main line sidings had been taken out of use by 1954 and were lifted in 1964. The Gipsy Lane quarry is still worked. The company manufactured paving setts, kerbs, window sills, spur stones, and broken stone for roads.

GROVE COLLIERY

Owned by Thomas Ball Troughton. Traded as Troughton & Lee. No. 1 colliery was working in 1841 but closed in 1844. No. 2 colliery was situated next to Bedworth station on the Bedworth side of the railway. There is no record of a rail connection, as it closed around 1851. There may have been a brickyard associated with this colliery, which Thomas Ball Troughton inherited from Charles Lilly of Coventry in 1834. Thomas Ball Troughton was Town Clerk of Coventry at one time. The colliery employed about twenty men in 1841.

THOMAS HARDY

Hardy had a brickyard in Chilvers Coton in 1828 and 1841 (he also traded as a timber merchant). He worked a brickyard in Stockingford by 1850, and was still trading in Chilvers Coton in 1866.

THOMAS HARRIS

Brick-maker in Nuneaton in 1841.

WILLIAM HARRIS

Brick-maker at the Bull Ring, Chilvers Coton in 1850.

HAUNCHWOOD BRICK & TILE CO. LTD

Formerly Haunchwood Brick Co. and initially Fowler & Knox. James Knox was passing through the new cutting of the Nuneaton–Whitacre Junction railway on his way towards Birmingham when he noticed the freshly exposed bands of Etruria Marl. With his trained civil engineer's eye, he knew this meant there was high-quality clay in the area.

This huge enterprise was originally founded as a partnership of George Fowler, Civil and Mining Engineer of Nottingham; William Maltby, Surgeon of Old Basford Nottingham; John Walter Fowler; and James Knox, Civil Engineer of Attleborough, Nuneaton. It traded under the name of the Haunchwood Brick Co. and was registered as a limited company on 11 May 1878 with directors James Knox, George Fowler, William Maltby, Robert Walker Smith, Charles Bennett, Richard Bush Spencer, and Herbert Walker. It had a share capital of £20,000.

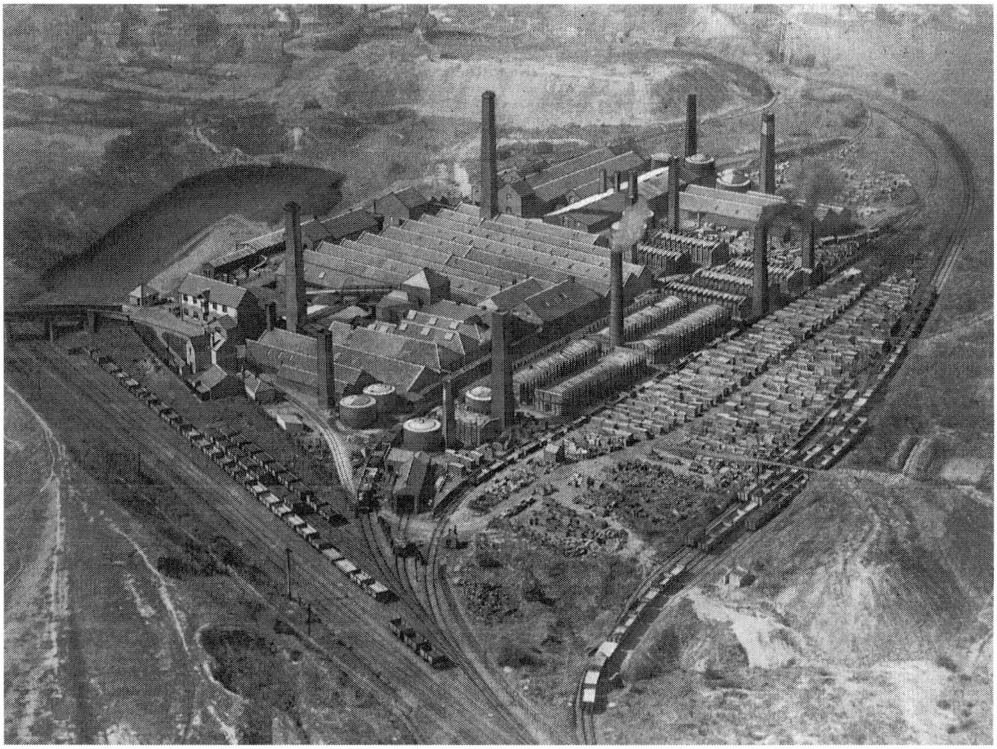

Above: Haunchwood Brick & Tile No. 1 yard, *c.* 1925. The main Nuneaton–Whitacre Junction–Birmingham railway line runs in the bottom of the photo. The Stockingford station in the Whitacre Junction is to the left. A loop line on the right led through to Haunchwood Nowells or Old Colliery, which closed around the time this photo was taken. Haunchwood Colliery's pithead buildings are hidden by the spoil heap at the top of the picture. Note the very extensive brick-stocking yards. The clay hole to the left and the footprint of these buildings lying hidden under a wooded area of open ground forming Whittleford Country Park, a local wildlife oasis. (Haunchwood Brick & Tile Co. Ltd)

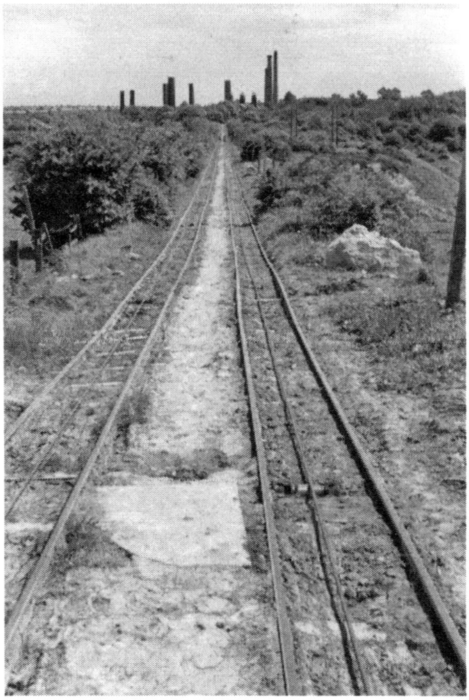

Right: The cable-hauled tramway transported rafts of steel trucks from the marl hole at Whittleford, Stockingford, back to Haunchwood No. 1 brickworks – some considerable distance, as can be seen. This view is from a point near the tunnel under Whittleford Road. Shunting and haulage was carried out at the brickyard's end (and No. 3 yard) with three Ruston Hornsby diesel-mechanical units.

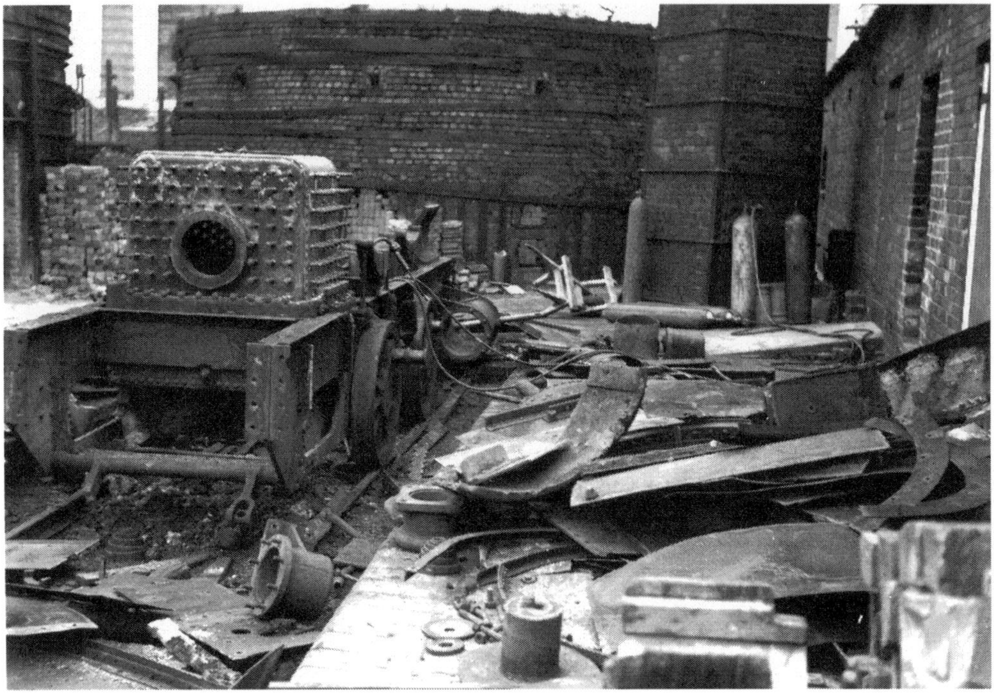

Above: The former Yorkshire Engine steam loco No. 2196 of 1927 bites the dust! Before Haunchwood bought it in 1930, it worked from new on a dam contract in Yorkshire. It is being dismantled here on site at No. 1 works by George Cohen Sons & Co. Ltd, scrap dealers, in January 1960. The works plate once used to reside on a mantelpiece in Stockingford. (Geoff Edmands)

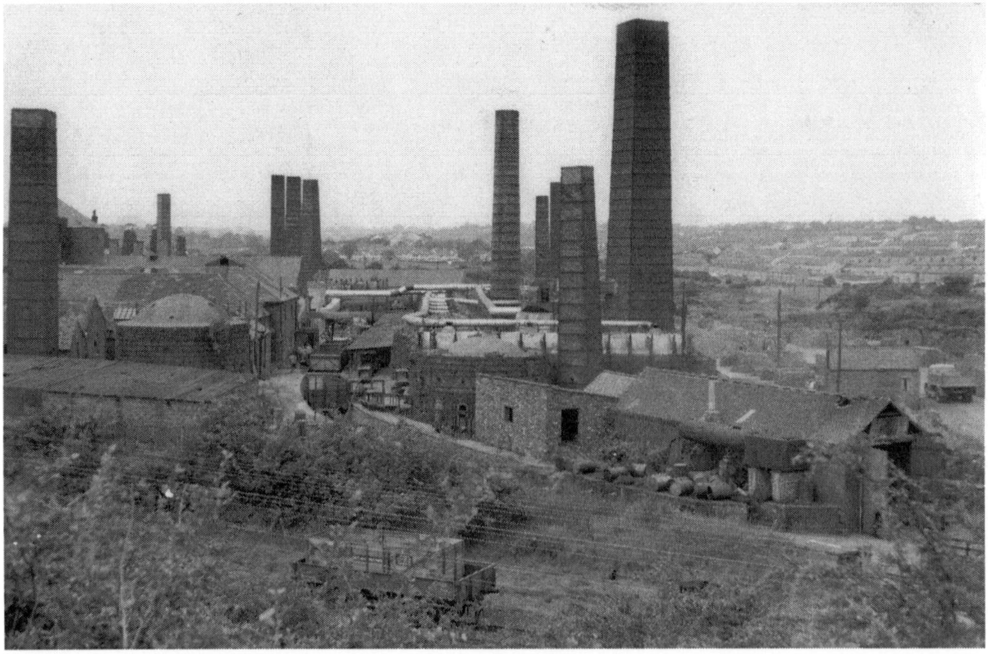

Opposite below: During the early 1960s, Haunchwood Brick & Tile spent considerable sums on modernising the works. A waste-heat recovery scheme was installed as part of this. You can see the metal ducts snaking around the roof of the kilns. This allowed recovery of hot air and gases from the brick kilns to be used in heating the dryers and the various brick sheds. The former steam engines that drove the machinery were dismantled and replaced with electric motors. The old steam locomotives were disposed of and a second-hand diesel purchased. This outlay only benefited the company for about eight years, as final closure came in October 1970. The loco shed on the right still shows telltale signs of a fire that broke out in 1928. It looks as though the hole in the roof was never repaired. (Geoff Edmands)

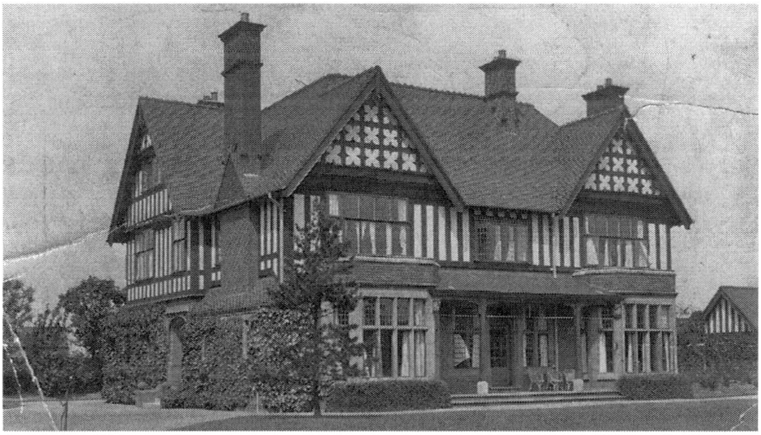

Above: James Knox had a substantial house built for himself and his family off Higham Lane, Nuneaton, called the Chase. It is a fine example of construction incorporating Haunchwood Brick & Tile products throughout. The building is now the Chase Hotel. (Mrs. E. C. Richardson)

The Knoxes were directors in other firms, such as Lewis Tileries, Barnstone Cement Co. Ltd, and Nottingham Patent Brick Co.

In 1896, the brickyard employed two men underground and one on the surface in a fireclay pit.

No. 1 works at Stockingford made the following: blue, brindled, and hard red engineering bricks (pressed and wire cuts), stable bricks and pavers, copings, ridge tiles and finials, chimney pots, flue linings, garden edging tiles, briquettes for fire places, sand-faced bricks, and garden vases.

In 1923, Haunchwood Brick & Tile's travelling salesmen were: Arthur J. Franks and his son Donald covering London; S. Wallace Allan covering Scotland; W. M. Hincks covering south-east England and East Anglia; J. H. Buxbaum covering the west of England; William Daffern, the north of England; Sidney Jones, Wales, Lincolnshire and Nottinghamshire; Fred Shute, the local area of Coventry, Nuneaton, Hinckley etc.; and H. T. Lavender, the Birmingham district. (Mr Lavender married the daughter of one of his Birmingham customers, and he then became a partner in the builder's merchant of Proctor & Lavender and bought millions of bricks from Haunchwood Brick & Tile Ltd. He was very well in with the local architects.)

Haunchwood exhibited at the Ideal Homes Exhibition at Olympia every year.

The associated company G. W. Lewis Tileries Ltd, whose head office was at Stockingford, had works in the Cannock and Essington areas of Staffordshire.

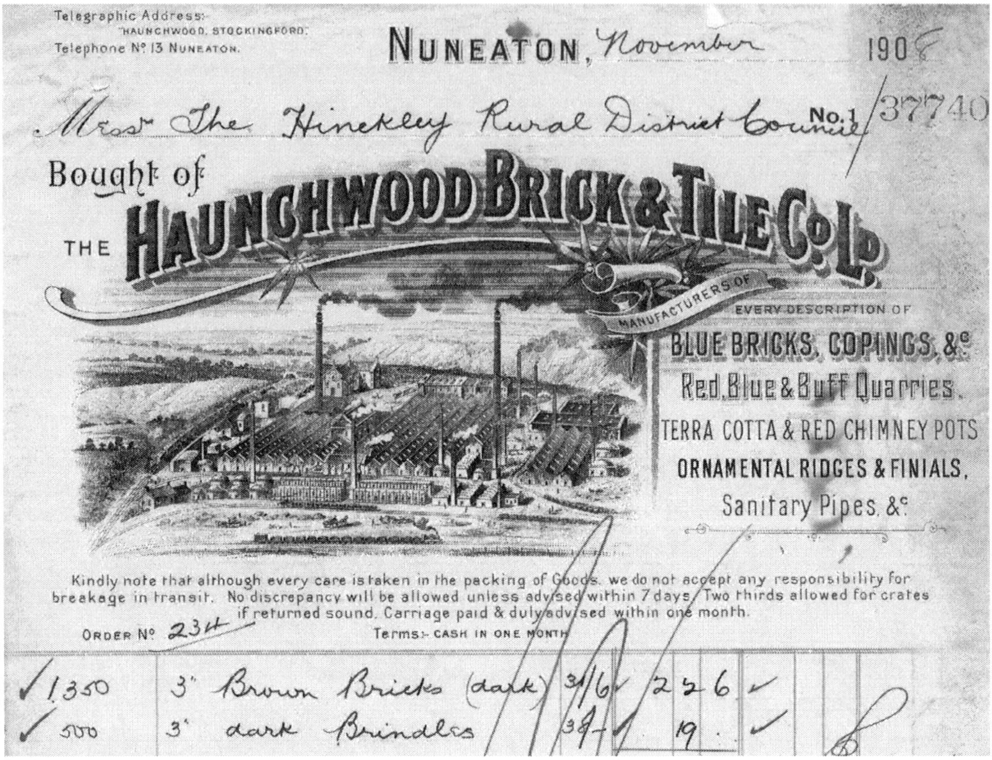

Telegraphic Address:- "HAUNCHWOOD, STOCKINGFORD."
Telephone No. 13 Nuneaton.

NUNEATON, November 1908

Messrs The Hinckley Rural District Council No. 1 / 37740

Bought of THE HAUNCHWOOD BRICK & TILE Co. Ld

MANUFACTURERS OF EVERY DESCRIPTION OF
BLUE BRICKS, COPINGS, &c
Red, Blue & Buff Quarries.
TERRA COTTA & RED CHIMNEY POTS
ORNAMENTAL RIDGES & FINIALS,
Sanitary Pipes, &c

Kindly note that although every care is taken in the packing of Goods, we do not accept any responsibility for breakage in transit. No discrepancy will be allowed unless advised within 7 days. Two thirds allowed for crates if returned sound. Carriage paid & duly advised within one month.

ORDER No. 234
Terms:- CASH IN ONE MONTH

✓ 1,350	3" Brown Bricks (dark)	31/6	2 2 6	✓
✓ 500	3" dark Brindles	38/-	19 .	✓

Haunchwood invoice.

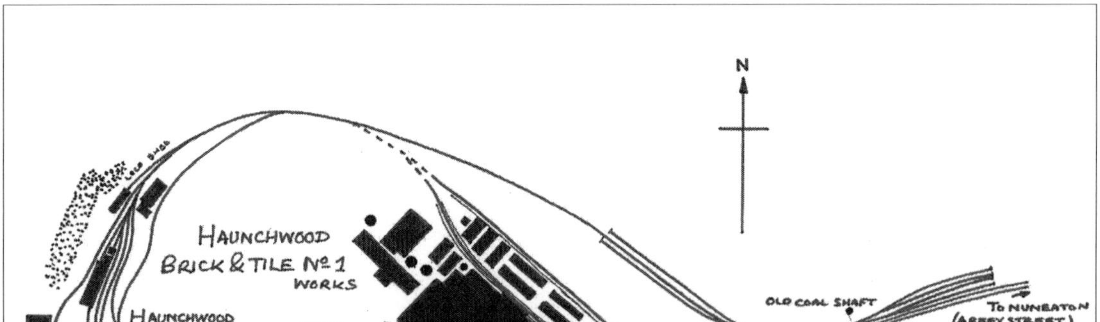

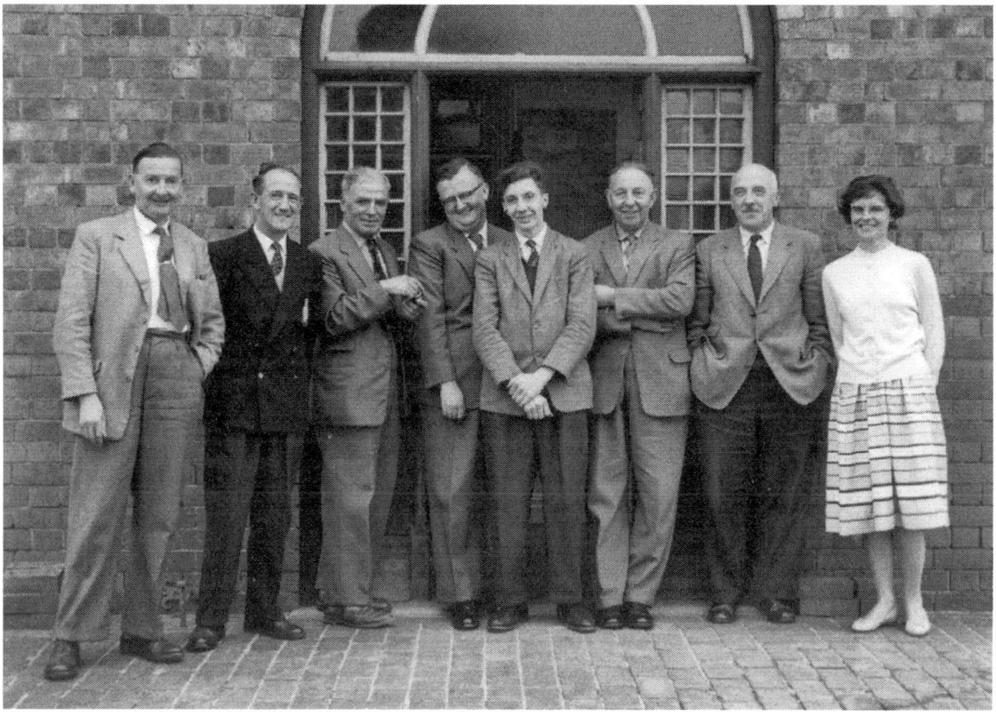

The office staff at Haunchwood Brick & Tile, 1964. From left to right: Geoff Edmands, Ernie Richardson, Frank Tonks, Dennis Portman, David Randle, A. L. Norton, H. Burdett, and Marie Cove. (Geoff Edmands)

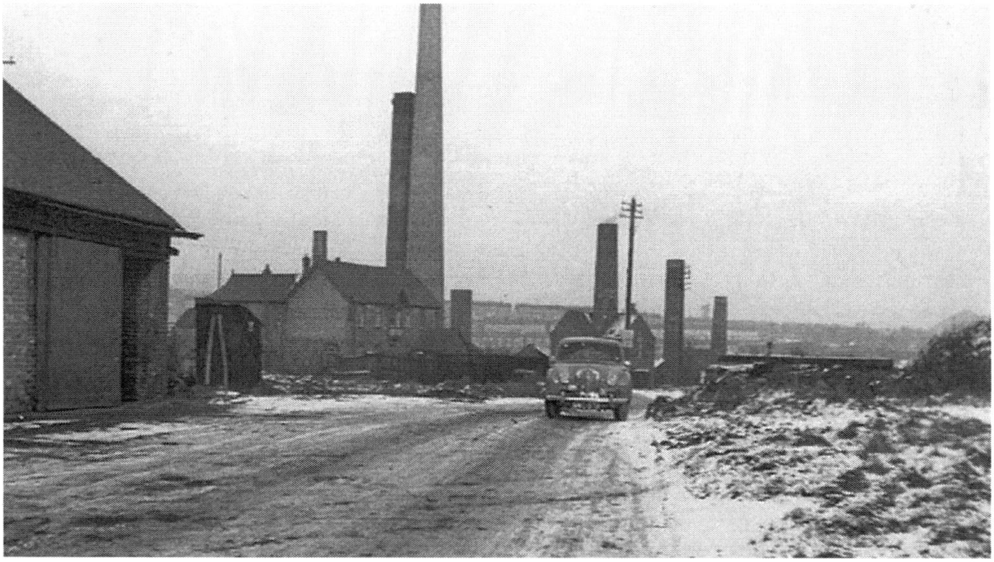

A winter's view of Haunchwood Brick & Tile, *c.* 1959, looking towards the offices just to the left of centre in the picture. The Camp Hill housing estate lines the horizon. (Geoff Edmands)

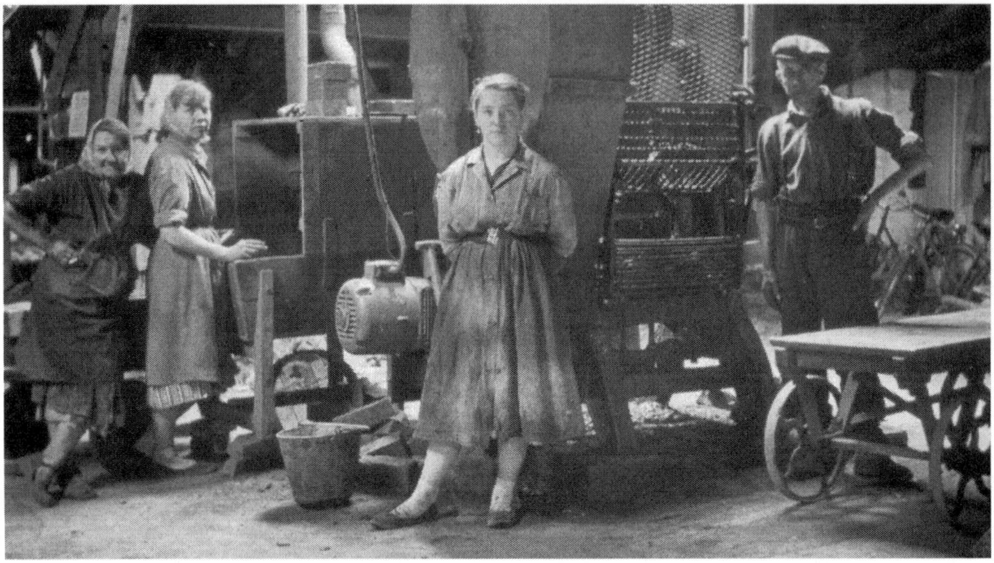

An evocative picture, taken on 26 April 1960, of ladies at Haunchwood Brick & Tile No. 2 brick press. From left to right: Mrs Hill, Mrs Thawley, and Margaret Wickenden in charge of a machine that was used to homogenise fresh clay. Mrs Thawley is standing with her hands adjacent to the square opening on the left of a machine that will improve the consistency of the clay prior to moulding. Mr H. Jeffcoat on the right will take the resultant piece of homogenous clay and trundle it away on the flat truck to be formed into bricks. The clay will be rammed and packed into a rectangular timber frame to form it into precisely the right size to be shaped into a block. Individual bricks were then cut with 'cheese cutter' wires (hence the brick term 'wire cuts'). (Geoff Edmands)

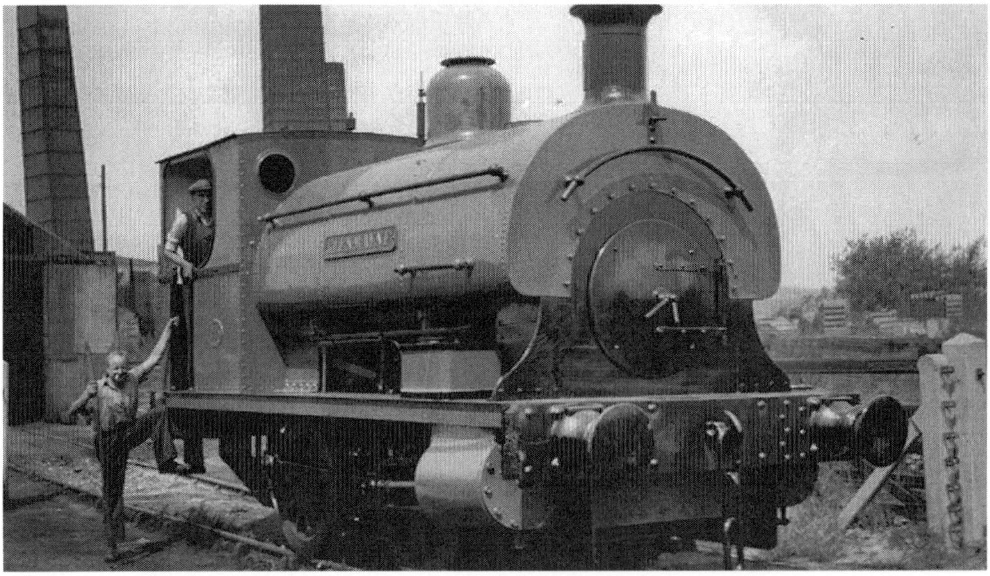

Haunchwood Brick & Tile's four-wheeled Peckett, *Glenmayne*, on 25 June 1952. It was acquired second-hand from Stanley Brothers when Charity Colliery closed. (Michael J. Lee)

By the 1930s, the two clay holes nearest to the brickyard were almost worked out and a new clay hole was developed in Whittleford Road. The former clay hole in Tomkinson Road was filled in by household rubbish by the Nuneaton Borough Council. In 1938, the brickyard was producing 10 million bricks per year.

Coal was purchased from Haunchwood Brick & Tile's sister company, Arley Colliery Co. Ltd, and salt for salt-glazing pipes was purchased from salt producers such as Amasol and the Salt Union Co. in Cheshire. This was brought by rail. Rail traffic ceased at the end of June 1967 and the track was lifted during 1969.
The company was merged with G. W. Lewis Tileries in 1968 and traded thereafter as Haunchwood-Lewis Brick & Tile Ltd.

No. 1 works closed in February 1971 due to dwindling clay reserves and impending smoke pollution legislation. 120 people were employed on closure. No. 3 yard employed seventy on closure. Production was switched to Essington near Cannock. Demolition of No. 1 yard commenced in 1971, but the footprint of the site and the old kiln bases

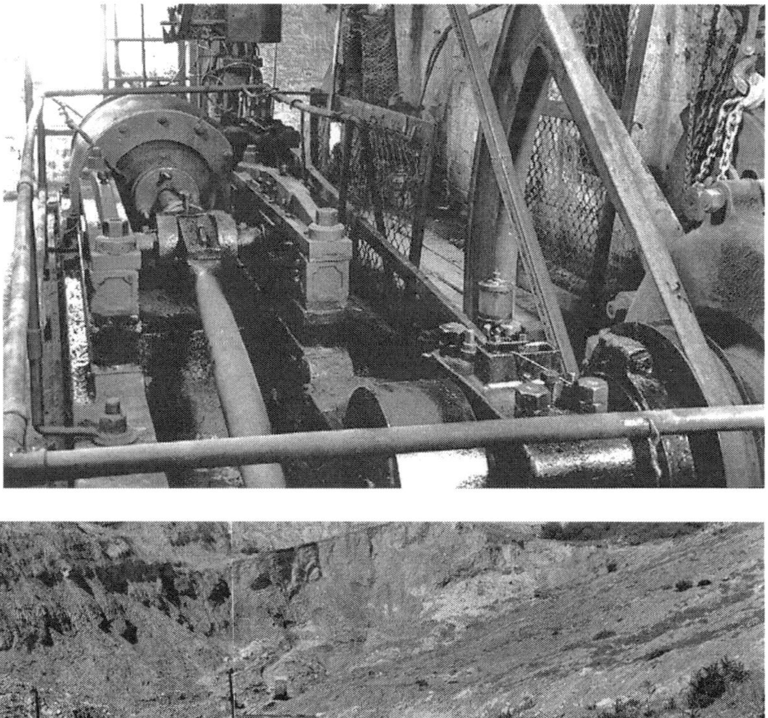

Haunchwood Brick & Tile had two of these stationary steam engines for driving internal machinery. This one was out of use by the time this photo was taken on 26 April 1960. It was replaced in January 1962 by an electric motor. (Geoff Edmands)

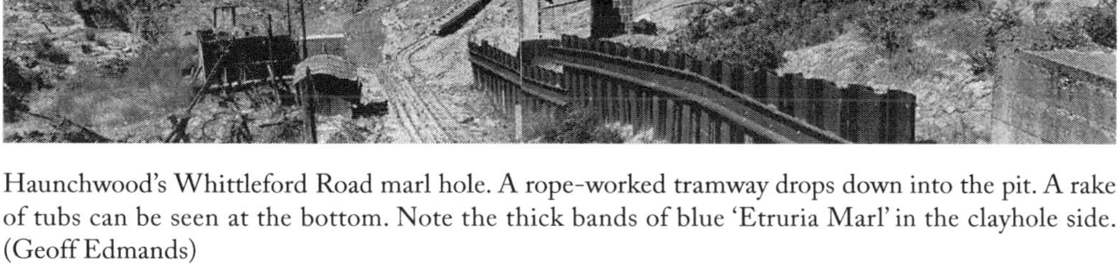

Haunchwood's Whittleford Road marl hole. A rope-worked tramway drops down into the pit. A rake of tubs can be seen at the bottom. Note the thick bands of blue 'Etruria Marl' in the clayhole side. (Geoff Edmands)

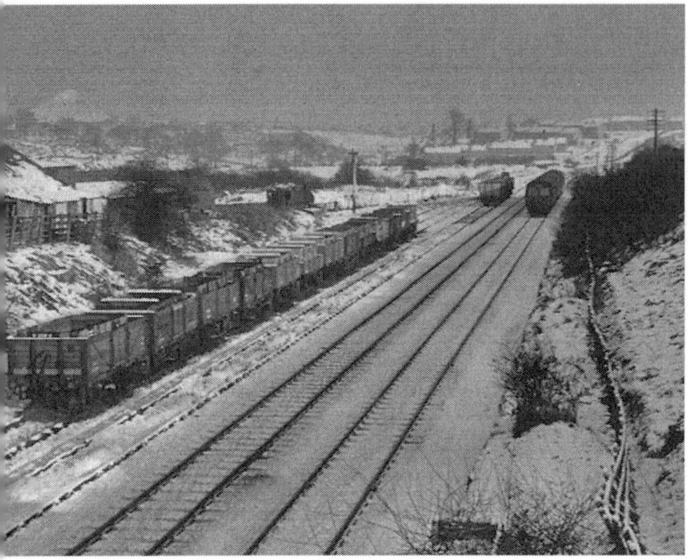

The interchange sidings of Haunchwood Brick & Tile and Nowells Colliery at Stockingford during the cold winter of January 1959. A long line of timber-sided wagons have been placed there for filling with outgoing clay goods. A shorter line of steel coal wagons is also present, used to deliver coal from local collieries to charge the kilns. In the centre of the picture, *Glenmayne* is in steam and has ventured out through the colliery/brickworks gates to engage in a modest amount of shunting. On the horizon is the conical waste tip known locally as 'Mount Jud'. A diesel multiple unit on the nearest line is heading towards Birmingham. We are looking back towards Nuneaton. (Geoff Edmands)

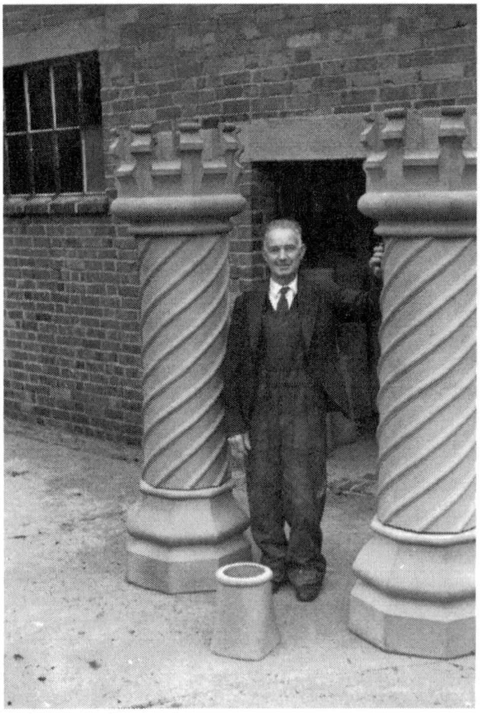

Samuel Preston, master chimney-pot maker, displays his finest artwork in these two elaborate chimneys, presumably for a large country house somewhere. The photograph was taken by Geoff Edmands on 2 June 1961 in No. 1 yard.

can be seen among the undergrowth of Whittleford Country Park. This wildlife haven is rich with memories of the old brick and tile industry of Stockingford, and it is being looked after by Friends of Whittleford Country Park. It is an absolute treasure.

Haunchwood Brick & Tile supplied all the bricks and flooring blocks for:

- Nuneaton town hall
- Hunterston power station in Scotland
- Nuneaton flour mill in 1885–86
- The railway cuttings approaching Euston station under contract to the London & North Western Railway
- Gillette Razor headquarters at Isleworth
- Barker's department store, Kensington High Street
- Derry & Toms, London
- D. H. Evans, Oxford Street
- Milk Marketing Board headquarters at Thames Ditton
- Earls Court Exhibition Building, London
- Swan & Edgar Ltd, Piccadilly Circus, London

- Peter Jones, Sloane Square, London
- Nottingham City Council House

Sand-faced briquettes were supplied to the ocean liner *Queen Mary* for use in one of her lounges. Blue platform copings were sold to Egyptian State Railways. Customers also included: the Patent Chimney Pot Co. of Leeds, Sankeys of Canning Town, the Fawcett Co., Diespeker & Co., and Kleine Patent Flooring Syndicate Ltd, London. Very detailed catalogues were produced by Cawthorne & Son, printers, Nuneaton. There was some inter-company trading between Haunchwood Brick & Tile and Stanley Brothers. Haunchwood had a stand at the Ideal Home Exhibition in the interwar period.

Most coal was supplied by Haunchwood (Old or Nowells) Colliery, Ansley Hall Coal & Iron Ltd, or Arley Colliery, which was owned by the directors of Haunchwood Brick & Tile Ltd.

No. 2 Works, Heath End, made red facing bricks, common red bricks and quarry tiles. Heath End Road closed in 1946.

No. 3 Works, Bermuda, made stoneware pipes and fittings, common bricks, hollow-fire clay blocks, and unglazed land drains. It closed in 1969/70. The business was taken over by Thomas Wragg of Swadlincote in 1970.

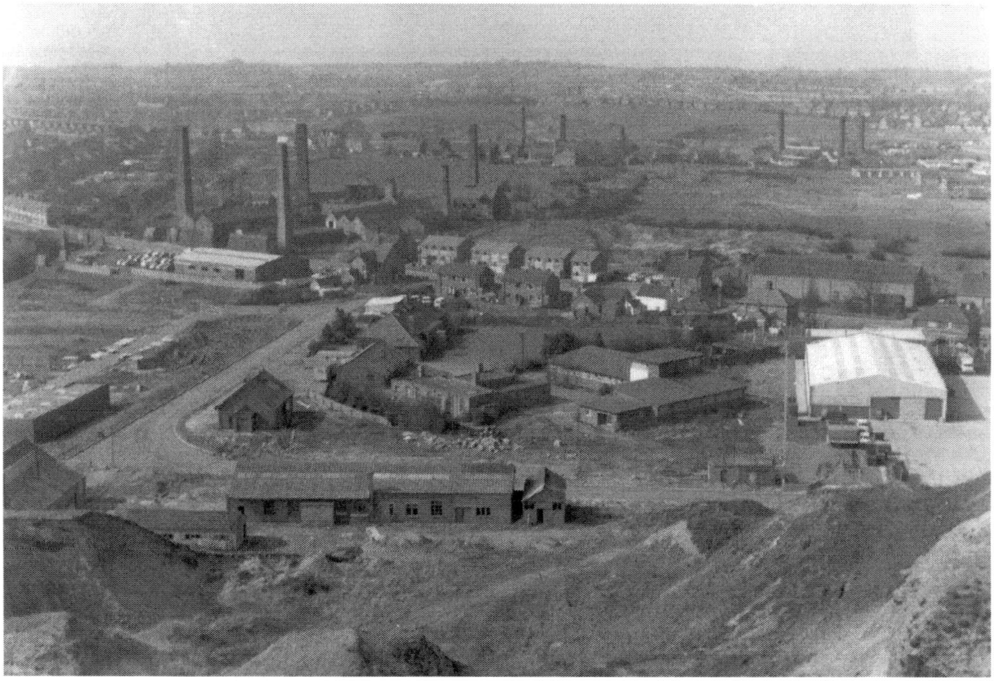

A view taken on 6 April 1968 from Griff No. 4's pit tip as extraction of shale was being carried out. The road running left to right is Heath End Road. The remains of Griff No. 4 yard is in the foreground. To the left, the cluster of three tall chimneys belongs to Haunchwood Brick & Tile's No. 2 yard, which was then about to be demolished, and in the distance can be see the Stanley Brothers yards still in production at this time. On the extreme left of the picture, the defile leading to a road underbridge is the Griff branch going towards Stockingford, where it formerly served Nuneaton Old Colliery and Stanley's brickyard. (Geoff Edmands)

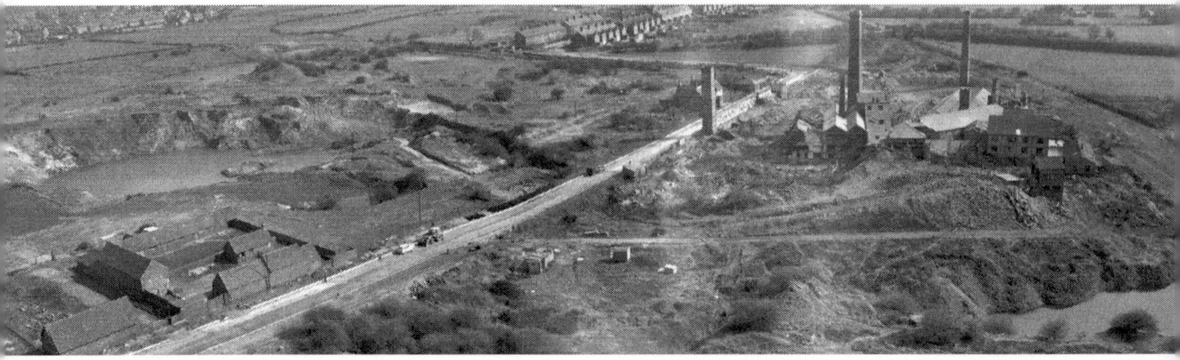

A view, taken on 6 April 1968, from the Griff No. 4 pit tip overlooking Bermuda Road with the derelict remains of Haunchwood Brick & Tile No. 3 yard on the right. The lifted trackbed of the Griff Branch railway curves in from the right. On the left can be seen the former clayhole used by the Haunchwood yard. In the centre top of the picture is the pit village of Bermuda, which served the new Griff Clara colliery when it was built in ten weeks in 1891. (Geoff Edmands)

HAUNCHWOOD (OLD OR NOWELLS) COLLIERY

List of proprietors:

- John Barber by 1770
- Alexander Donald (1745–1808)
- William Dorsey (son in law of Alexander Donald)
- Peter Unger Williams (operated Charity Colliery in Bedworth and a brickyard on Nuneaton Common) (1775–1837)
- Caroline Williams (Peter Unger Williams' widow) (1780–1873)
- John Nowell (active *c.* 1852–1881)
- Sir Alfred Hickman (operated in the period 1882–1925 until final closure)

The earliest reference to coal being dug in the Haunchwood area was in 1353, when a complaint was brought against John de Allesworth for digging in the prioress's ground on 'le Hanche' at Nuneaton and taking marsh coal to the value of 10*s*.

In 1801, a horse-drawn tramway was laid, connecting the colliery with the Coventry Canal.

Nowell & Son were in charge by 1852. John Nowell sunk a new shaft by 1859 to the Two Yard Coal. In 1882, 200 acres were let with coal undug. In 1891, 220 men were employed. In 1896, the colliery employed 362 underground and 89 on the surface. It worked the Slate coal. There were 538 men employed in 1910. In 1913, the figure was 417 men; coal output was 148,000 tons. In 1920, 364 men were employed. On closure in 1925, 431 were employed. A large customer of this colliery was the adjacent Haunchwood brick yard.

On closure, one of the shafts was fitted with a pump for reducing the amount of water seeping into the nearby Tunnel colliery workings. On closure of Tunnel pit, the pump was switched off and the pithead was finally demolished in 1968, the work being completed by 19 July 1968.

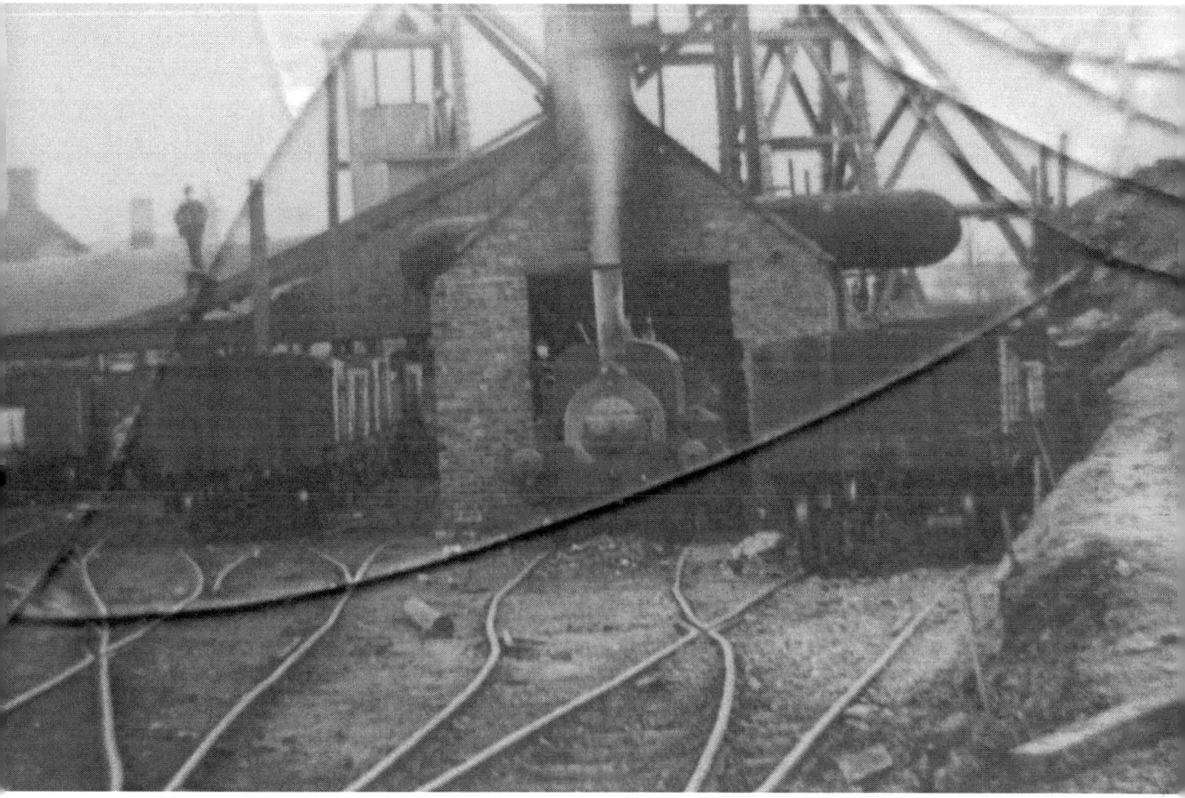

Local enthusiasts were stunned to discover this photograph of a hitherto unknown loco at Haunchwood Nowells Colliery despite the poor condition of the glass plate. The original was given to me by Ian Hickman, whose grandfather was Sir Alfred Hickman, who took Haunchwood over in 1882. Industrial railway expert the late Mick Lee studied it carefully and believed it to be a 'Black Hawthorn' loco that had 12- or 13-inch cylinders. There is no record in the Warwickshire colliery or IRS records of this engine, but it is possible that it was a loco called *Alfred* that Sir Alfred Hickman used at his Bilston Furnaces site (BH882 of 1886). Although its subsequent history at Bilston is not known, it could have been transferred to Haunchwood as necessity dictated. After the passage of much time, we will never know for sure. The lightly laid flat-bottom track, spiked direct to the poorly aligned sleepers, suggests that there must have been plenty of derailments. Wooden baulks and sprags lie around, ready to be placed in front of wagon wheels to stop them rolling. IRS records at that time listed Haunchwood Colliery locos Nos 3 and 4 and not Nos 1 and 2, so it seems likely there were two earlier locos and this could have been one of them. The other continues to intrigue industrial railway enthusiasts locally. (Hickman family)

HAUNCHWOOD (TUNNEL) COLLIERY

This was officially opened on 20 July 1891 by Sir Alfred Hickman on the site of Woodland Cottage Farm. Sinking was finally completed on 1 January 1894. The chief shaft sinker here was Mr Elijah Hyde.

A brickworks was also operated primarily for the manufacture of bricks used in the construction of the colliery, but later for production of common bricks. The brickworks closed in 1901.

In 1896, the colliery employed 498 underground and 110 on the surface. It worked the Two Yard coal. There were 1,426 men employed in 1910, 950 men in 1945, 680 in 1950, and 1,238 in 1957. In that year, 391,700 tons of coal were produced. In 1960, 946 men were employed below ground and 294 on the surface.

A private platform was provided here for the miners' 'Paddy' train, which came into use by May 1922. The Paddy ran from Abbey Street station, where a couple of old Midland Railway coaches were kept for this specific purpose. After being used by miners in their 'pit black', they were unsuited for anything else. The Paddy was reduced to one coach by the 1950s, and this was burnt out in a spectacular fire one night, fuelled by all the coal dust embedded in the seats.

Haunchwood Tunnel colliery officially closed on 25 March 1967. The pithead buildings were taken over for other purposes, and some still remain as part of an industrial estate.

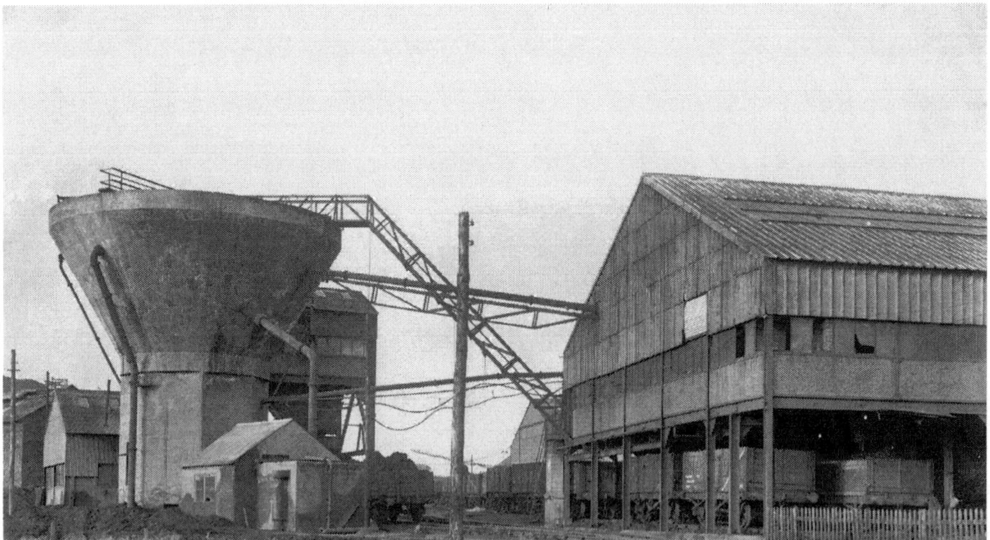

The screens and washery of Haunchwood (Tunnel) Colliery, 1946 (Horace Dudley Studios)

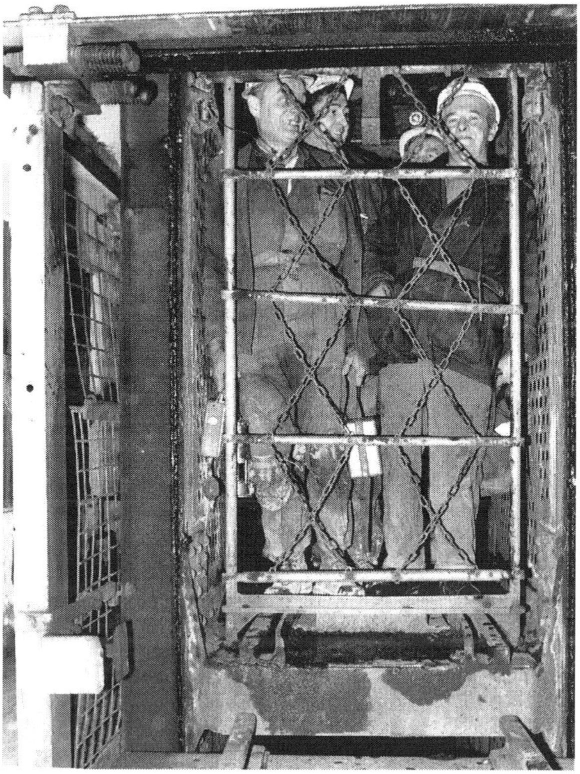

Right: Descending in the cage at Haunchwood. (Author's collection)

Below: The bottom of the shaft at Haunchwood Colliery, 1946. (Horace Dudley Studios)

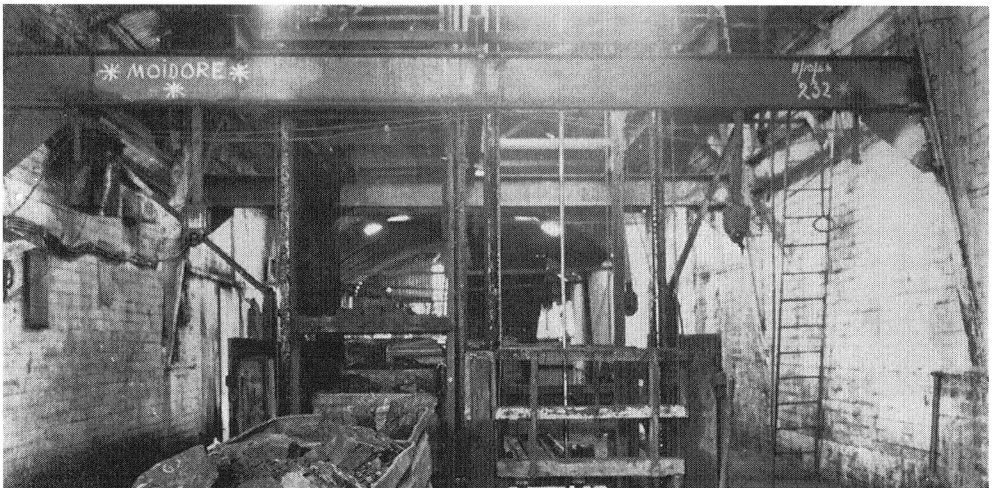

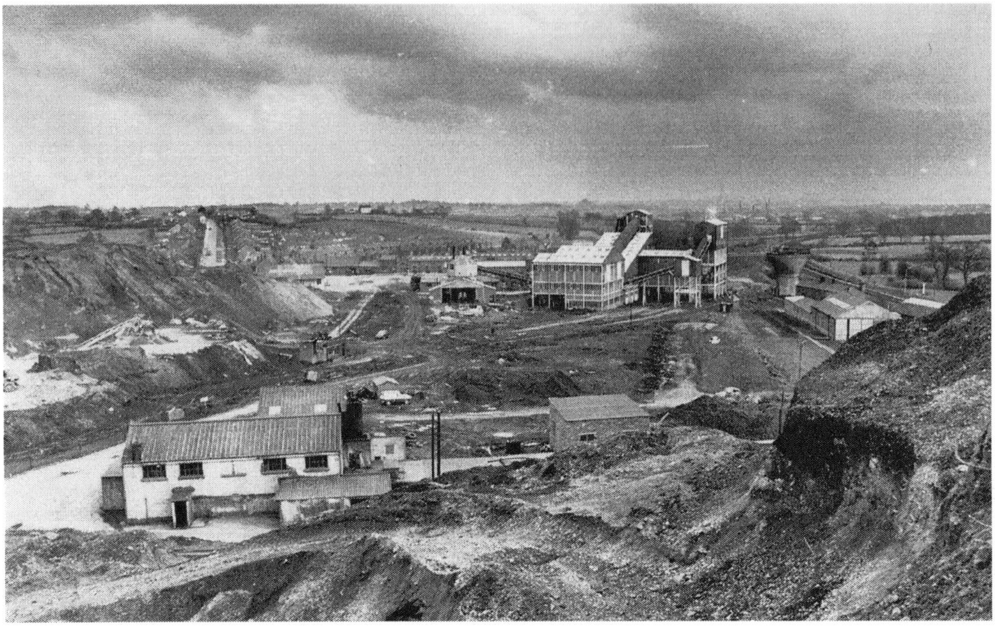

This is a view taken sometime around the time of closure at Haunchwood Tunnel Colliery. The photographer has climbed the pit tip to record the scene.

A BR Type 2 diesel hauls a stone train towards Birmingham, *c.* 1961, as it heads towards Stockingford tunnel. Haunchwood Tunnel Colliery is in the background. To the left is the small volcano that was the colliery tip. You would be hard pressed to know today that the tip was once here so well has it been transformed, landscaped and smoothed over. (Mike Mensing)

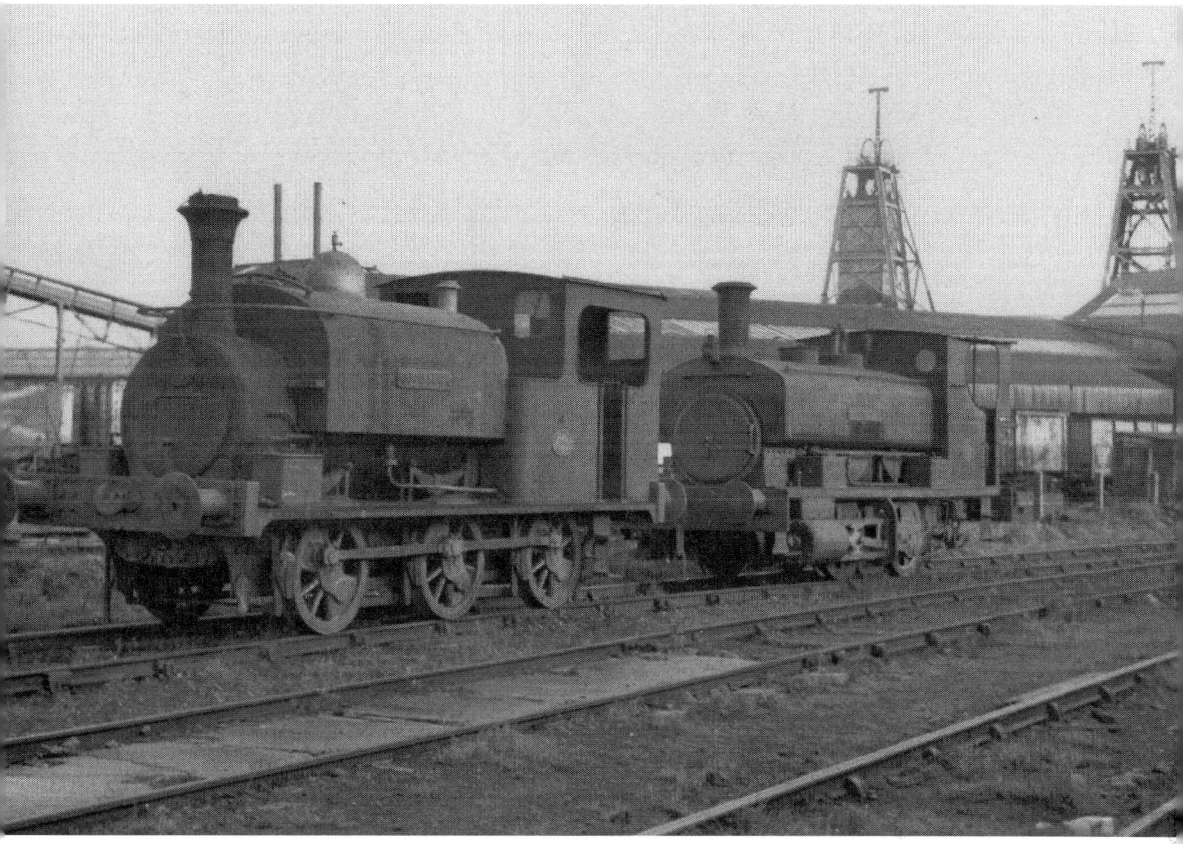

Two former Griff Colliery locos: *Good Luck*, nearest the camera, and *Success* were transferred to Tunnel colliery and both were scrapped on site by J. Cashmore Ltd of Great Bridge, Staffordshire, in June 1966. *Success* was transferred away from Griff on 21 December 1960. *Good Luck* came in September 1962.

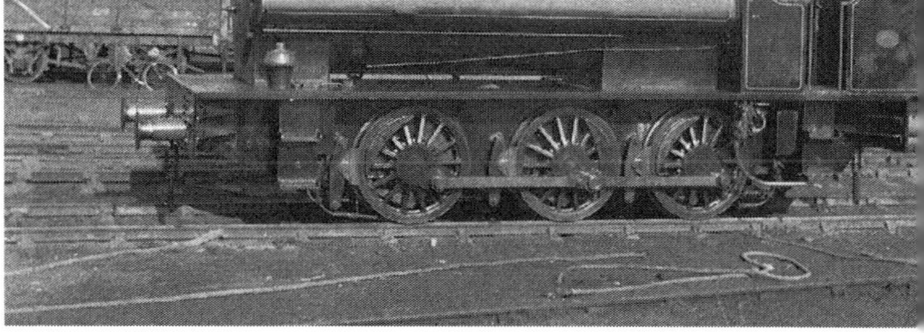

Haunchwood Colliery No. 1 was a relative latecomer to Tunnel pit. New in 1954, it was built to the standard Austerity design by the Hunslet Engine Company of Leeds and was a powerful addition to the colliery stud. Sadly, it had a very short life and the National Coal Board did not get the best value from her. She was scrapped by John Cashmore Ltd of Great Bridge, Staffordshire, in October 1967.

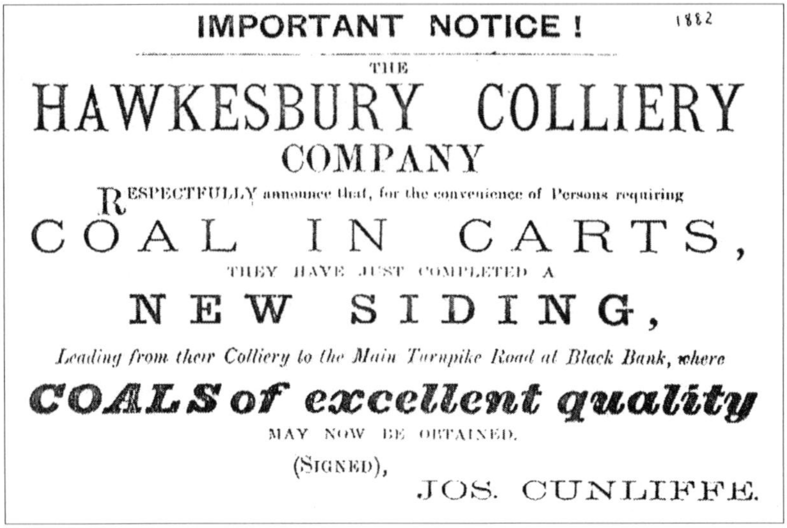

Hawkesbury advert.

HAUNCHWOOD WOOD PIT

This was an outlying shaft within the ancient Haunch wood. It was probably associated with Haunchwood (Nowells) Colliery. But apart from the name on a map, nothing else is known about its operation.

HAWKESBURY AND BEDWORTH BRICK & TILE CO. LTD

Benjamin Pollard Blockley was the proprietor in 1910. It was taken over by Blockley's Ltd of Hadley, Shropshire, around 1932.

The brickyard was built on Charity's Nos 3 and 4 collieries' site. A siding existed for delivery of coal from Charity Colliery close by, and a private siding connected to the Coventry–Nuneaton line, which was installed in 1897. The branch to Charity Colliery closed with the pit in 1924, although the Brick & Tile complex retained its rail connection until 1948–52. The brickworks' address was Marston Lane, Bedworth.

HAWKESBURY COLLIERY

Owned by Parratt, Fernyhough and Whieldon in 1776. George Whieldon was the owner in the 1830s. The early shafts close by were slate coal pit, new engine pit, bye pit, Heal's pit.

Mining was being carried out as early as 1682 from lands leased from the Coventry Corporation.

A Newcomen engine had been installed here in 1725. A Boulton & Watt engine was purchased 1776. An iron furnace was erected near Hawkesbury Colliery about 1840; the iron was found to be of a good quality, but production only lasted a year due to insufficient quantity of ore to make production viable. Iron ore was dug widely locally and later shipped to Staffordshire for processing. In January 1886, a new colliery siding was opened, giving access from the colliery to the main turnpike road at the Black Bank, Bedworth, for shunting wagons of coal for taking away by cart traffic. There was a stoppage due to ingress of water in nearby Exhall Colliery in March 1887, but mining was resumed after a few weeks in April that year.

The colliery closed on 18 July 1888, with the loss of 300 jobs. A new pit at Hawkesbury had been opened. Known as 'New Winnings', it came into use in 1860 and was operational until 1948.

HEATH, CHADWICK & CO.

(see Nuneaton Colliery)

THOMAS HINCKS (1776–1843)

The brickyard of Thomas Hincks is recorded in 1845 as being operating after his death by his widow, Mary. It was on the site of Atack's billiard hall in Regent Street, then called Brick Kiln Lane. It stretched back across the later course of the Trent Valley main railway line. Thomas Hincks (1776–1843) was also proprietor of the town's flour mill. The yard and kilns were obliterated by the construction of the railway. No tramway is indicated. There was a similar brickyard nearby, also on the bed of the TVR, but the operator of this yard is not recorded. (See Peacock Lane brickyard.)

JEE'S QUARRIES & BRICKWORKS

It traded as R. R. Jee until 1877. It was later Jee's Hartshill Granite & Brick Co. Ltd. Tarmac operated the site from 1 September 1972, but quarrying ceased by 2000.

These became extensive workings, originally started by Richard Jee, a local landowner at Hartshill from modest beginnings in 1822.

The quarry tramway was operated by horses up to the workfaces, by rope haulage up to the sides of the quarry, but on the surface locomotive haulage was used. The gauge of the track was 2 feet 6½ inches. Locos were extensively used for overburden removal. Three steam locos were obtained from the Hunslet Engine Company between 1907 and 1910. There were three second-hand engines from Bagnalls of Stafford between 1913 and 1931. A further Hunslet was purchased

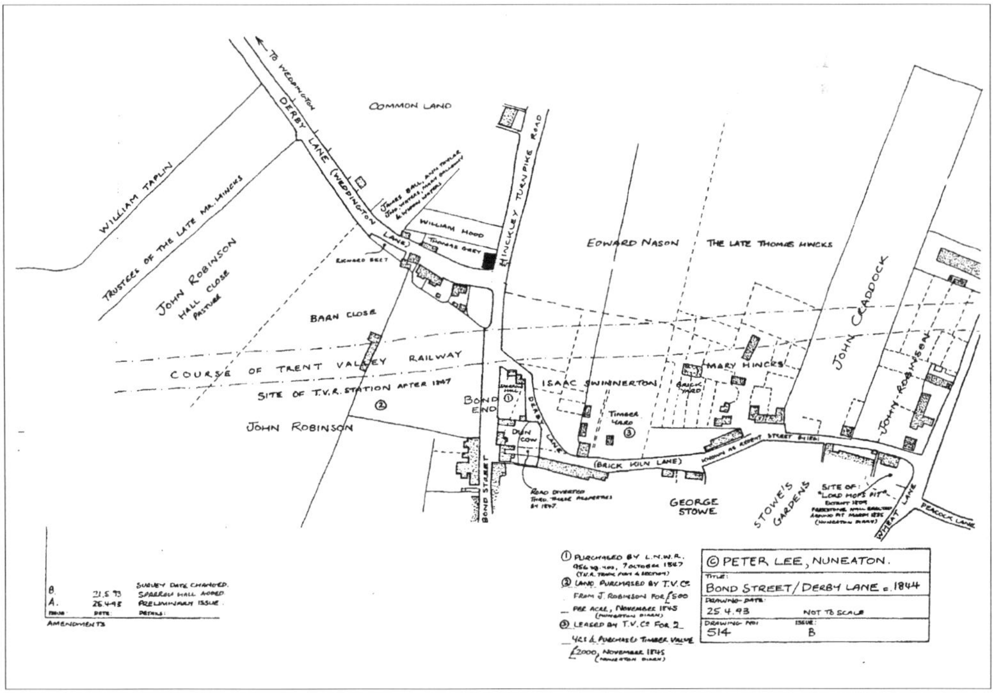

The site of Thomas Hinck's brickyard was in Brick Kiln Lane, Nuneaton (Regent Street today).

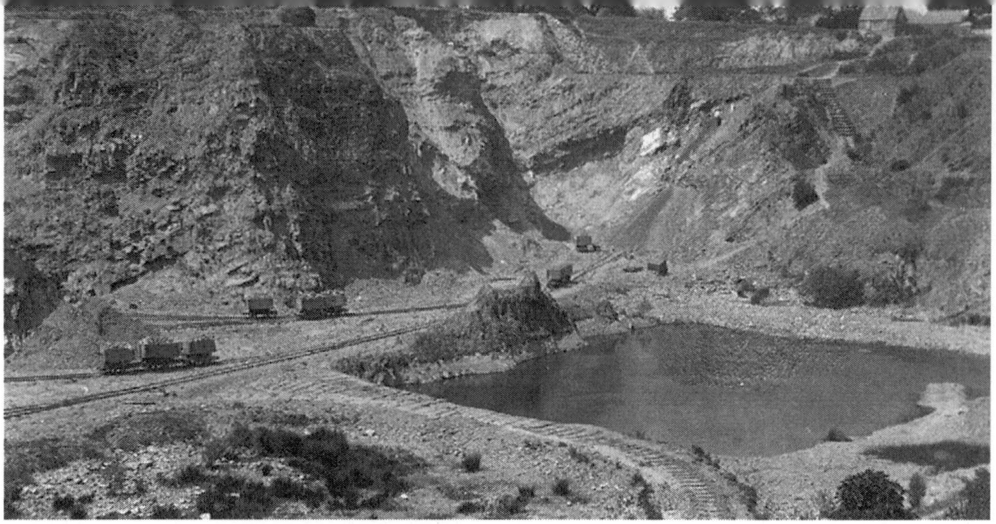

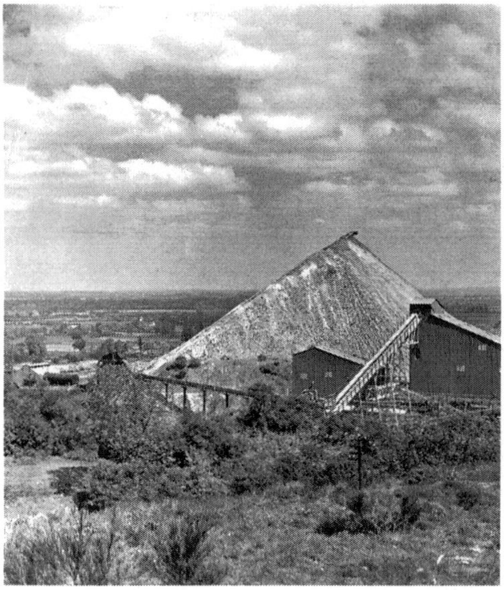

Above: Jees Quarry in the early 1900s, with small laden tubs of stone ready for haulage to the lip of the quarry, and then over the trackage down to the railway sidings or the canal wharfs. (Jean Lapworth collection)

Left: A panoramic view of Jees Quarry in the 1960s. The countryside in the distance is Leicestershire. (Author's collection)

second-hand from a contractor in Ireland, in addition to an ex-WD 4-6-0T Hunslet of the type widely used behind the front lines in France in the First World War. These were gradually replaced over the years by seven Ruston diesels. The narrow-gauge tramways were subsequently replaced by mobile plant and were taken out of use by October 1954.

The last steamer on site, a 0-6-0T Bagnall named *Stafford*, narrowly missed preservation when it was inspected by enthusiasts in its shed in June 1954 with preservation on the Talyllyn Railway in Wales in mind. They went back the following week to conclude the deal and unfortunately found it cut up. One of the diesels, a four-wheel diesel mechanical built by Ruston & Hornsby in 1940, did make it into preservation in March 1957 and can be seen today on the Talyllyn Railway under the guise of *Midlander*. The last three diesels, which entered service between 1942 and 1951, were sold for further use in 1955 to a customer in Scandinavia. The rest of the diesels were scrapped on site by 1960.

A private siding existed, allowing stone to be removed by rail over the national network. The siding went out of use around 1990.

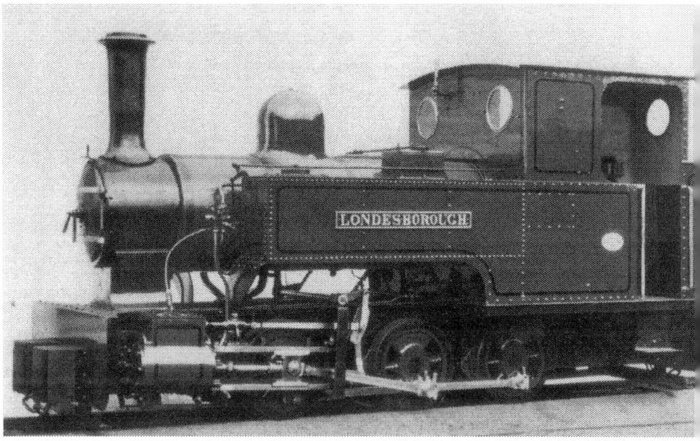

Jees purchased *Londesborough* new in 1931 (works No. 2449). She was a very modern-looking steam loco, delivered to the quarry in the maker's standard livery of green-lined red and yellow. She was cut up in 1951. A diesel loco was purchased to replace the steamer, and was also named *Londesborough*. Its replacement diesel, along with two others from Jees, was sold to a company in Sweden. Their subsequent fate is not known. (Official makers' photo from William Bagnall Ltd. Collection of Maurice Billington)

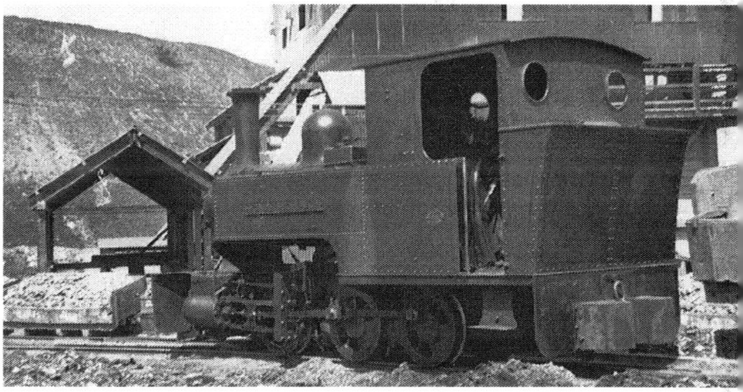

Londesborough at work at Jees, with various items of stone-processing plant in the background (K. J. Cooper, IRS)

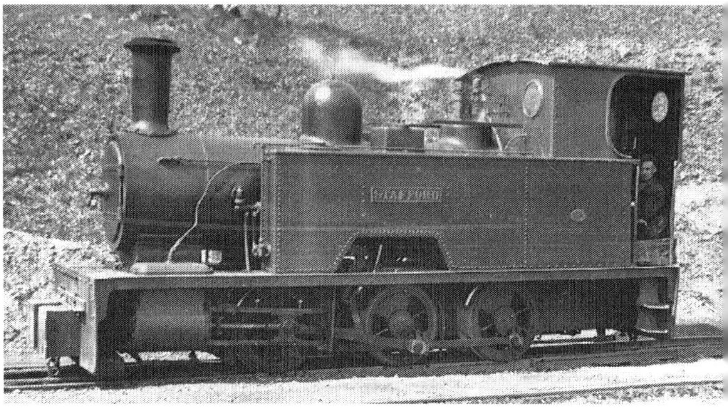

Stafford was another Bagnall but this time was purchased second-hand from the Admiralty (works No. 1911 of 1917). It had worked when new at W. D. Neasden. Jees purchased the engine from the Ridham Salvage Dept at Sittingbourne in Kent, where some of the First World War second-hand material was sent for disposal. Abandonment of the quarry railway system came in October 1954. In June 1954, *Stafford* was the subject of a preservation attempt. Several local members of the Talyllyn Railway Preservation Society went to view *Stafford* and make contact with the demolition contractors. After initial enquiries, they went back the following week to meet the contractor's manager, only to find *Stafford* cut into pieces. For some years this engine was engaged on working spoil trains on an overburden line that crossed main Mancetter Road on a steel bridge to a tip nearby. When visited in 1946, it was painted in dark green with red lining out. *Stafford* was the last steamer in use and would have made a very useful engine had she survived into preservation. (M. J. Lee)

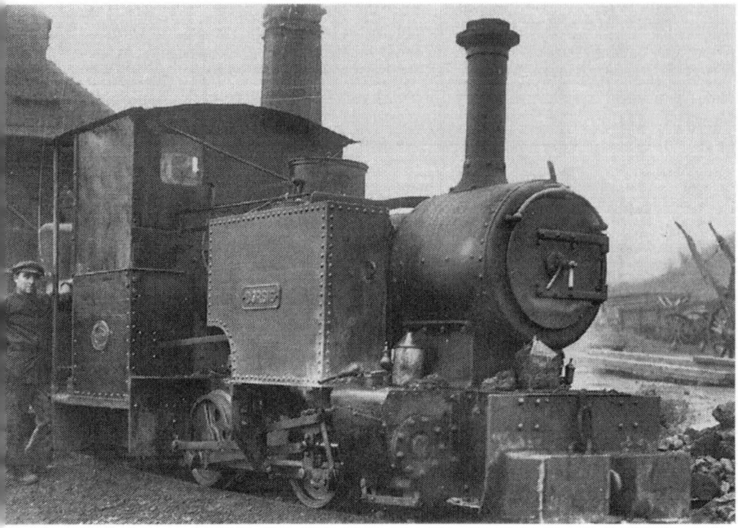

Dorsib (named for Doris and Sybil, the quarry manager's daughters, I think) was a queer-looking engine. It gives the impression of having been cobbled together from spare parts, but it was in fact built new, a peculiar-looking side tank by Hunslet. Jees purchased it in 1910 (their works No. 1039). This view, taken in the early 1930s, shows it just prior to disposal, as it is thought to have been scrapped around 1935, by which time it was probably thoroughly worn out. Jees purchased three new Ruston Hornsby diesels around this time to replace some of its aged steamers. (Real Photographs/IRS)

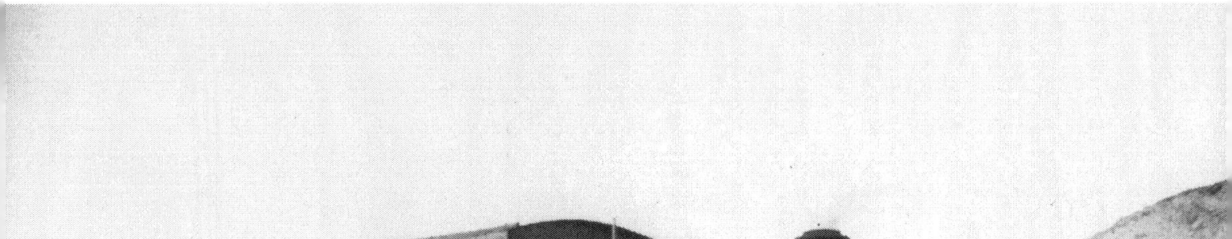

Butcher provides a lift for Jees workers as it trundles along sometime in the 1930s. *Butcher* was one of the typical quarry loco products by Bagnalls of Stafford (works No. 1655 of 1901). The enclosed cab was probably fabricated at the quarry. Footplate work in some exposed areas of the quarry must have been a bleak job in winter. She was scrapped in 1945. (IRS)

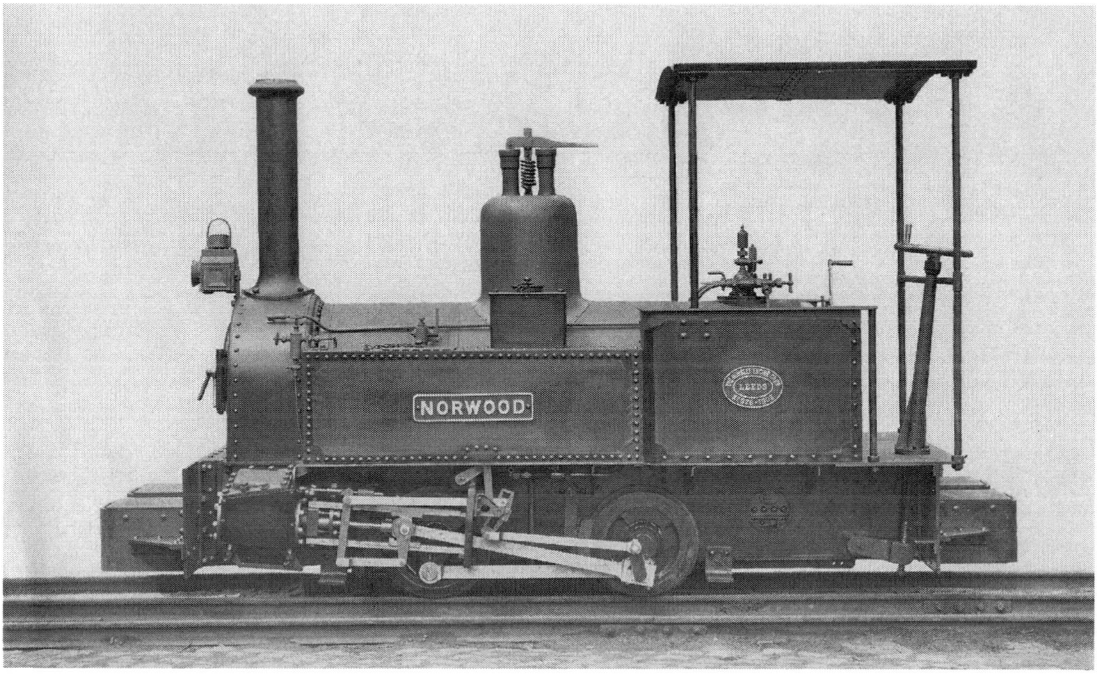

Hunslet's built many quarry locos, but they were usually fitted with a saddle tank as standard. *Norwood* was different. Here it is in ex-works condition, built as a side tank. Sold in 1908 as works No. 976, it was scrapped about 1935/36 and its nameplate transferred to a new Ruston Hornsby diesel delivered in 1936. (The Hunslet Engine Co. Ltd/Maurice Billington collection)

MANCETTER GRANITE QUARRIES

This was owned by Eli Green by 1876, then Thomas H. Green by 1891, Ireland & Knight by 1891 until about 1908, then W. L. Ireland until about 1932. Thereafter, it was known as Mancetter Granite Quarries Ltd until 1951.

The quarry appears to have started in 1876 and was originally known as 'Reservoir Quarry', as it adjoined the Oldbury canal reservoir. The Stockingford branch of the MR was opened throughout in 1876, and it seems likely that Eli Green carted his stone to the head of that branch, where he despatched his material thence by rail at a wharf there. The head of the branch for many years was known as 'Green's Siding'.

In June 1891, T. H. Green entered into a private siding arrangement with the LNWR on the Trent Valley main line. The first Ordnance Survey shows a tramway to the canal as well by 1887. A narrow-gauge railway connected the workings to the LNWR siding, and steam locos were purchased. The railway system went out of use in April 1956 and subsequently the rails were dismantled. The quarry continued in use until the 1990s.

MOORWOOD BRICKWORKS

Owned by Jee & Co. Ltd. A brickyard was operated as an offshoot of Jees Quarries and was served by a spur from the Stockingford branch at Chapel End. The connection was established in the 1890s.

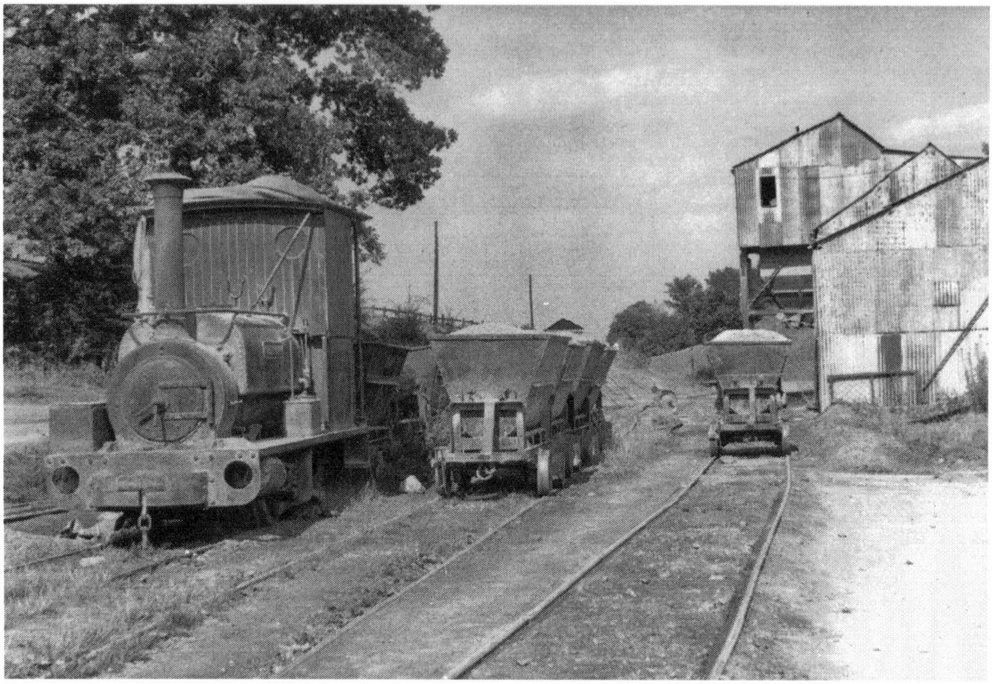

Mancetter Quarries had a couple of these standard Hunslet Quarry locos. *Oldbury* was No. 754 of 1901. The first loco, *Mancetter*, had disappeared by the early 1950s. *Oldbury* was lying derelict beside the crushing plant at the quarry at this time. She had acquired timber side sheets to the cab that was improvised by the quarry workshops. By now the stone traffic was being worked by a Ruston Hornsby diesel. *Oldbury* was scrapped on site by Asher Brothers, Codnor, Derbyshire, in February 1955. (Maurice Billington)

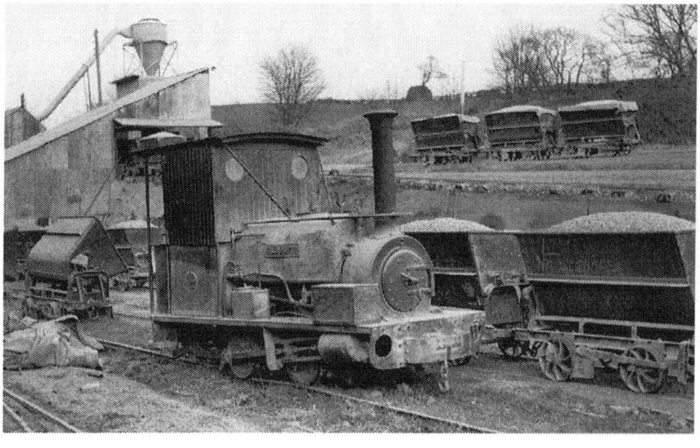

Oldbury occupies some very dubious trackwork in this view. If the rails here were typical of local quarries, derailments would have been an everyday occurrence. This picture gives a good impression of the quarry tubs in use at that time. (Maurice Billington)

Opposite, top left: Mancetter, as built new in a works photograph by Hunslet Engine, No. 598 of 1893. It lasted into the 1930s or 1940s, but its precise disposal particulars are not known. (Hunslet Engine Co. courtesy of Maurice Billington)

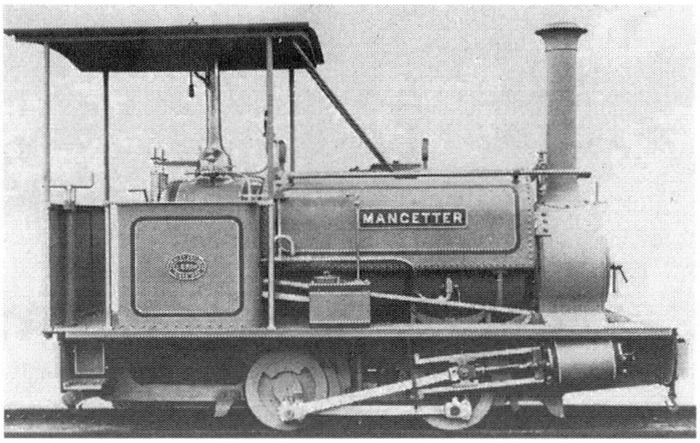

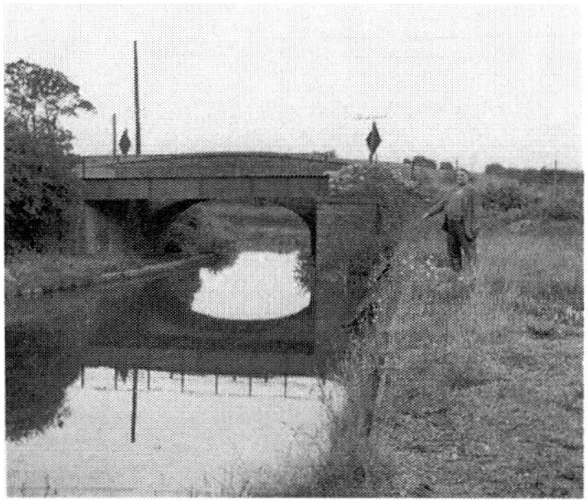

Above right: The overgrown remains of the wharf on the Coventry Canal. The tramway ran down to the railway wharf nearby via the steel bridge in the background, which was built alongside the road bridge. The rotund gentleman was pointing out an abandoned quarry truck that had slipped into the canal and been left there long after the rails were removed. The photo was taken in the early 1950s. (Maurice Billington)

Brick-making on this site ceased in 1912. The brickworks' chimney stood isolated for many years and was demolished in the 1950s.

JUDKINS QUARRY

William Cropper & John Judkins were in partnership from the mid-1840s until 1862. The quarry was then owned by William Arthur Judkins & Co. until 1903. (The leasehold was purchased outright from the Aston Estates in 1864.) It became a limited company in 1903 and traded as Judkins Ltd.

Judkins Ltd took over the former Boon's Windmill Hill Quarry in 1955 and that quarry was absorbed within the greater workings of Judkins and was obliterated by their workings. In 1970, Amey Roadstone Corporation took over the quarry and it traded as ARC from 1972 onwards.

The original quarry was small in 1817 and was recorded then as being just 1 acre in size. By 1860, it reached over 7 acres in extent. After 1864, further acquisitions of land took place, expanding the quarry to 48.5 acres. In 1892, 45 acres were rented on land leased from James Tomkinson of Willington Hall, Tarporley. A small brickyard was also worked. The bricks had the initials W. A. J. impressed in the frogs. Brick-making ceased about 1890. The clay pit was filled in with quarry waste and the site was then leased to the National Telephone Company for the manufacture of concrete underground telephone ducts. There was a canal dock on the Coventry Canal adjacent to the quarry. A narrow-gauge railway was built in 1864, connecting the quarry faces to the adjacent Nuneaton–Birmingham and Leicester line; it was opened by 1870/71. The gauge of the tramway was 36 inches. There were two steam locomotives acquired in 1911 and 1912. Prior to that time, horses were used for haulage. These steamers were joined in 1930/31 by petrol locomotives.

Up until 1906, the offices consisted of a stone-built structure overlooking the canal. This was demolished in October 1964. Offices fronting Tuttle Hill were built for the quarry engineer in 1933; they later became the main office block in 1939. In 1935–36, a macadam plant was built to provide tar paving and asphalt for road building. A subsidiary was formed: Modern Asphalt Pavements Ltd for tarmac contracting.

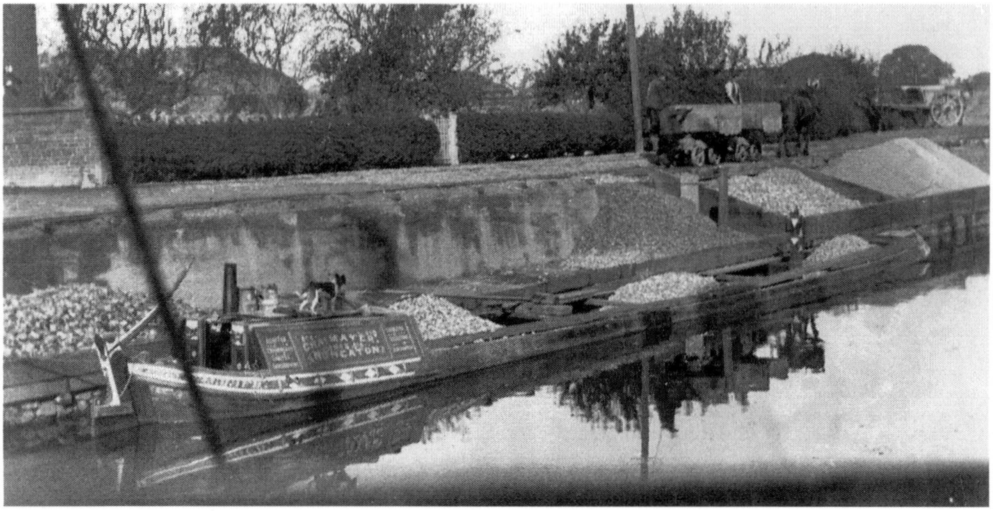

Judkins wharf on the Coventry Canal, *c.* 1900. The photo illustrates three ways of transporting the product. A narrow-gauge tramway brings stone to the wharf. A short, horse-drawn train of two wagons is just to the right of the telegraph pole and unloading is in progress. A road cart can just be seen on the extreme right of the picture. Bins of different stone grades are arranged adjacent to the canal so as to be convenient for loading. A boat, owned by Charles Mayer, is laden with stone ready for despatch. Mr Mayer was also a director of the quarry company and part owner of the Punch Bowl pub. (Author's collection)

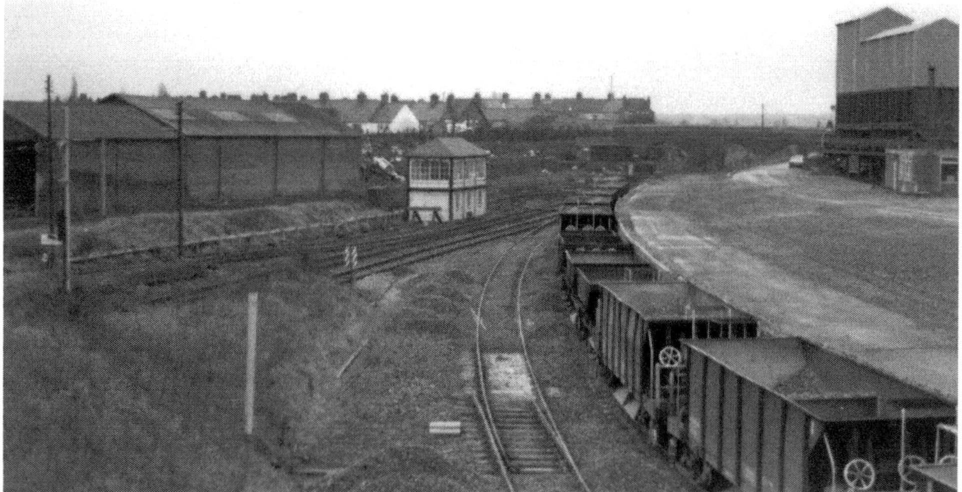

Judkins stone wharf at Nuneaton Midland Junction. Abbey Street station was beyond the overbridge that carried the main Nuneaton–Birmingham road over the railway at this point. The trackage in the foreground was once double track and formed the former Ashby & Nuneaton joint line in the direction of Market Bosworth, which was served by the Midland Railway only at this point and jointly with the London & North Western Railway from Weddington Junction. To the left of the signal box is the MR goods shed. Stone-hopper wagons are being filled on the right. Judkins stone-crushing plant can be seen. (Author's collection)

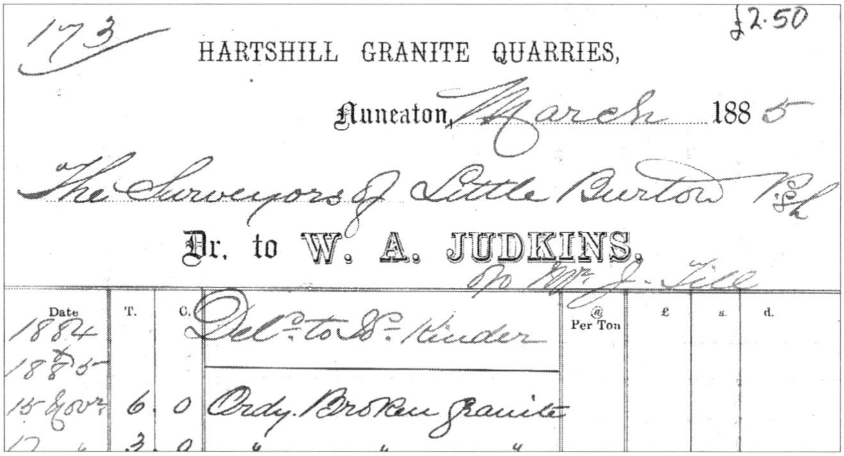

173 £2.50

HARTSHILL GRANITE QUARRIES,

Nuneaton, March 1885

The Surveyors of Little Burton

Dr. to W. A. JUDKINS,

Date	T.	C.		@ Per Ton	£	s.	d.
1884			Del. to Kinder				
1885							
15 Nov	6	0	Ordy. Broken granite				

Judkins invoice.

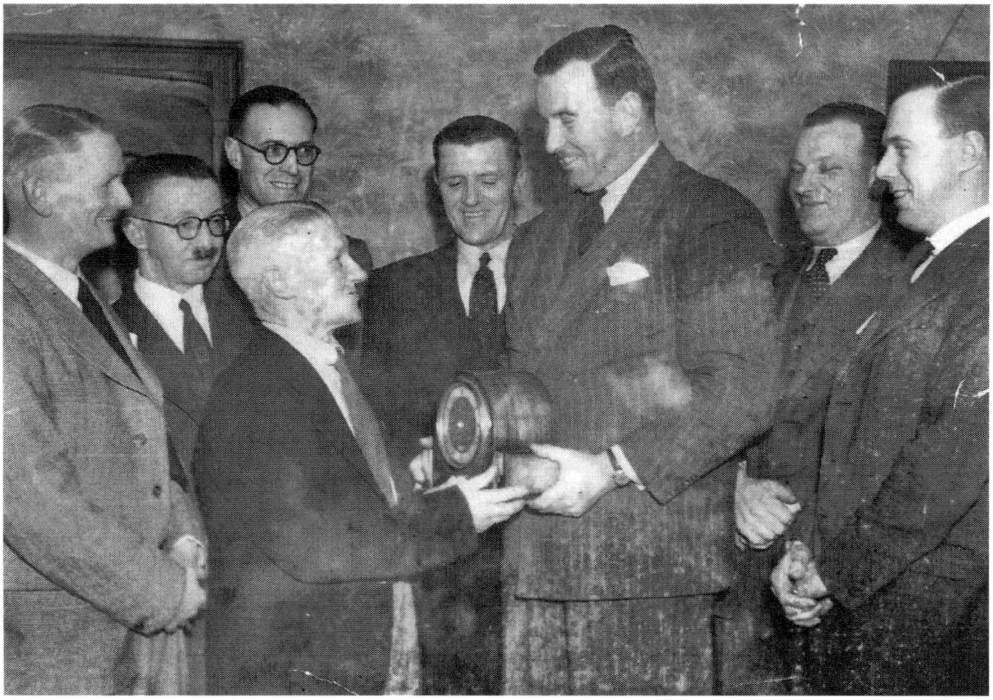

A photograph records a great day in the life of John (Jack) Jeffcoate in January 1949, when he celebrated fifty-five years' service with the company. The presentation was presided over by Mr C. W. Hopkins, quarry manager, and the clock was presented by Mr Myrton Judkins, managing director of the company. In his fiftieth year with the company, Mr Jeffcoate received a gold watch and a medal of long service from the Institute of Quarrying. After his retirement, he remained with the company as a consultant and advisor so that the business could continue to benefit from his great wealth of experience. The presentation was accompanied by a musical soirée given by Mr Alan Fenn, vocalist, and Mr W. Haines on the piano. The presentation took place at the Nags Head public house in Queens Road. (Author's collection)

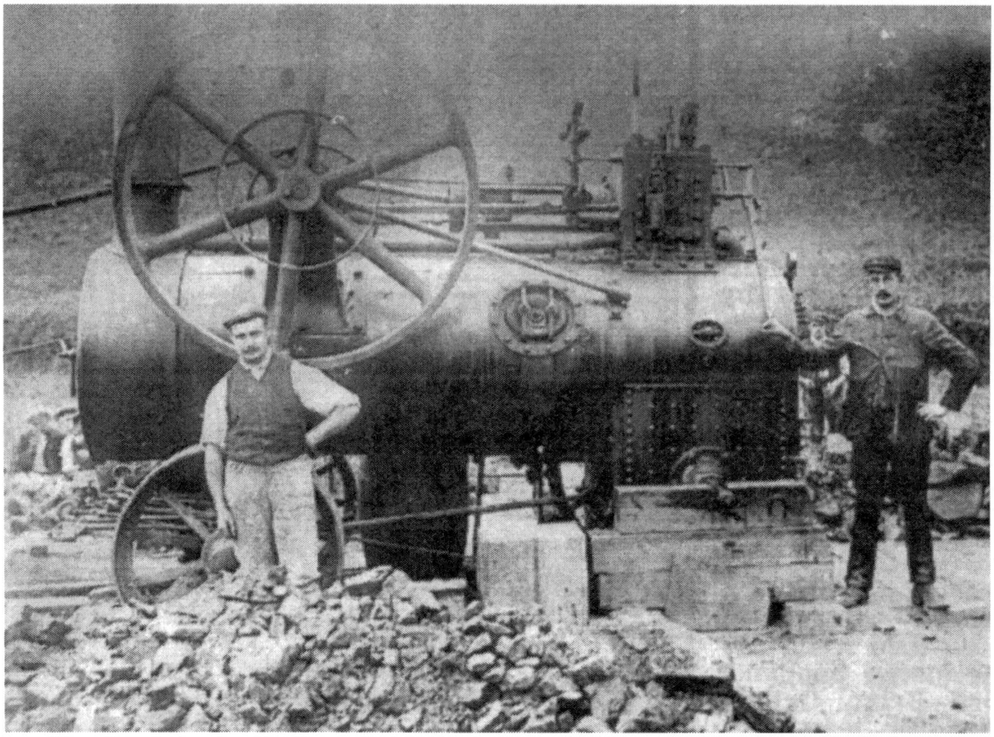

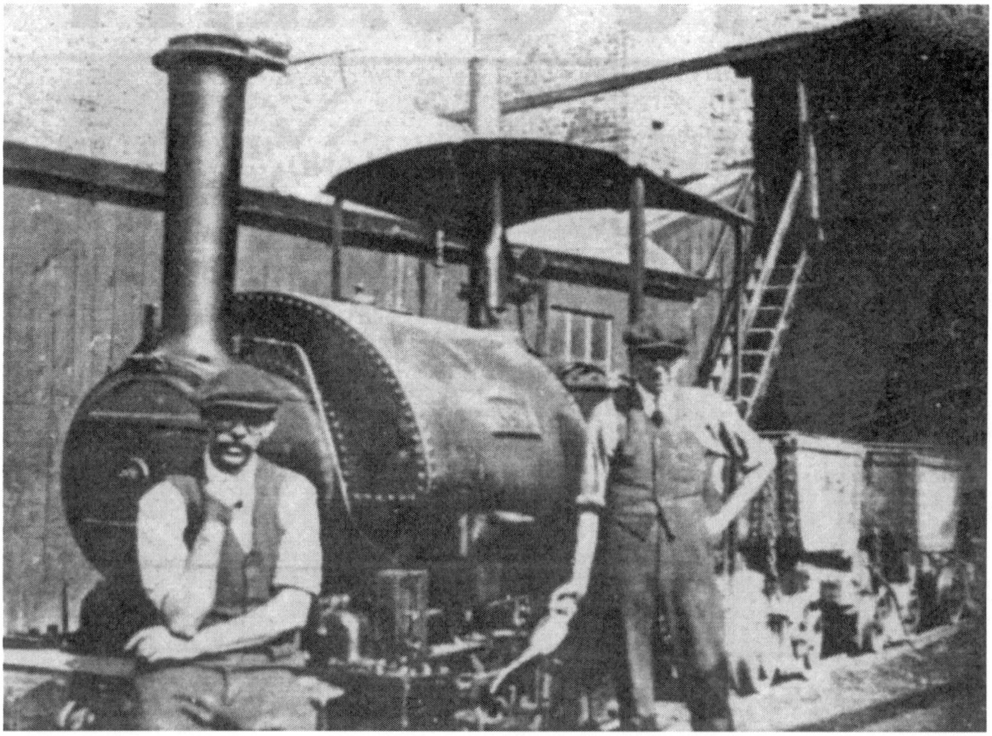

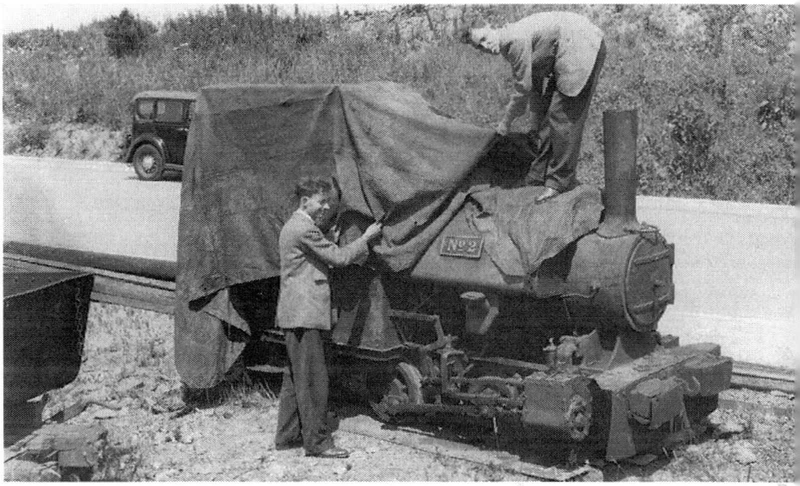

After the abandonment of the railway system, the track was dismantled and the Bagnall loco No. 2 was laid aside with a tarpaulin over it. One of my correspondents, visiting the site in the late 1940s, was offered the entire narrow-gauge stock and equipment for the cost of removing it from site. Unfortunately, my contact had just been demobbed from the RAF and, living in a flat, found the prospect of this donation enticing but completely impractical. These were the days, of course, before wide-scale rail preservation, so the logistics of setting up a narrow-gauge railway was virtually unheard of. No. 2 here is being unwrapped for inspection by members of a railway society. A little later it was scrapped on site by Oswald Bond, a dealer from Cardiff, around 1949 – probably not long after this photo was taken. No. 1 on the other hand had a happier fate. It was sold on to M. E. Engineering Ltd of Cricklewood in London in 1956 and afterwards went to British Railway's Beeston Creosote Works, where it survived to be rescued, I believe, by my correspondent, who had been offered the railway in the 1940s. The loco possibly still resides in his garden somewhere in Oxfordshire. (K. J. Cooper, IRS)

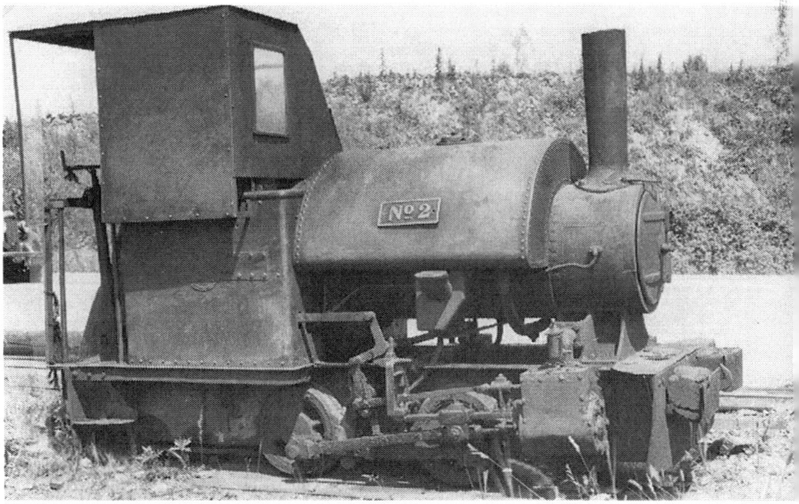

No. 2 fully exposed for the benefit of photography. The curious half-cab had been fabricated to replace the original standard maker's cab and was designed to pass through a part of the works with very limited clearances. (K. J. Cooper, IRS)

Opposite above: Chief shot-firer Tom Pinfold (right) and his crewman with the converted traction engine used for stone-crushing at Judkins quarry around the turn of the twentieth century.

Opposite below: George Pickard Snr (left) and Harry Pinfold tending the Bagnall steam loco, which appeared on the scene in 1911.

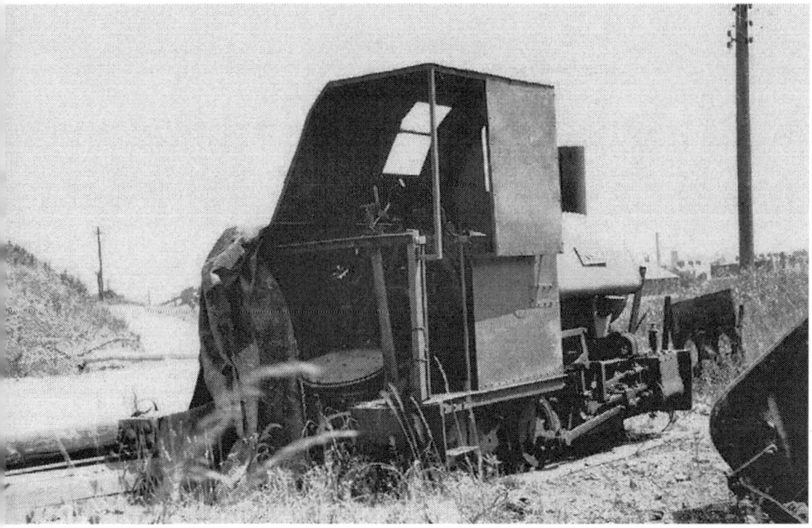

A rear view of No. 2 and its rickety cab. In this picture a few derelict skip wagons are scattered around too. (K. J. Cooper, IRS)

Another company, East Cambrian Gardens Ltd, was formed in 1946 for laying out tennis courts, bowling greens, and ornamental garden work. This latter company's trading style was changed to Judkins (Coated Macadam) Ltd in 1950. Its name was changed again in 1964 to Judkins (Civil Engineering) Ltd. A subcontractor called Hawker's of Smethwick provided road transport from 1934. Gradually, Hawker's transport was superseded by Judkins' own lorries after the Second World War. The Hartshill quarry of Man-Abell Quarries Ltd was purchased in 1962.

MARSTON HALL, PERCH HILL & PINK HILL QUARRIES, MARSTON JABBETT

Owned by Camwell & Boon (1860), B. J. Forder & Son Ltd (1904, 1924), then known as London Brick Co. & Forders Ltd, and latterly London Brick Co. Ltd.

The quarries at Marston Jabbett were begun by 1850. They were small and worked out by the 1920s. The London Brick Co. Ltd took over the site and removed the crushing plant and conveyors to provide a brick-distribution depot. A railway siding branch from the Trent Valley main line had been opened on 27 March 1904. After the stone trade had ceased, the branch was used to bring lengthy trains of bricks from the LBC's Bedfordshire brickworks for local distribution by lorry. They used three petrol- and diesel-engine shunting locomotives for handling the internal shunting.

A narrow-gauge tramway had formerly been used around the quarries, worked by horses, but the track was abandoned after stone extraction ceased.

MIDLAND STONE QUARRIES LTD

Located at Tuttle Hill, Nuneaton, this quarry was opened in 1880 on agricultural land owned by a farmer called David Ensor, who found digging stone more profitable than farming the land. A further small pit was opened later. This was connected by a narrow-gauge railway. The tramways were either rope-worked or drawn by horses; they were dispensed with in the 1950s in favour of road transport. The quarry has recently been extensively stabilised and readied for

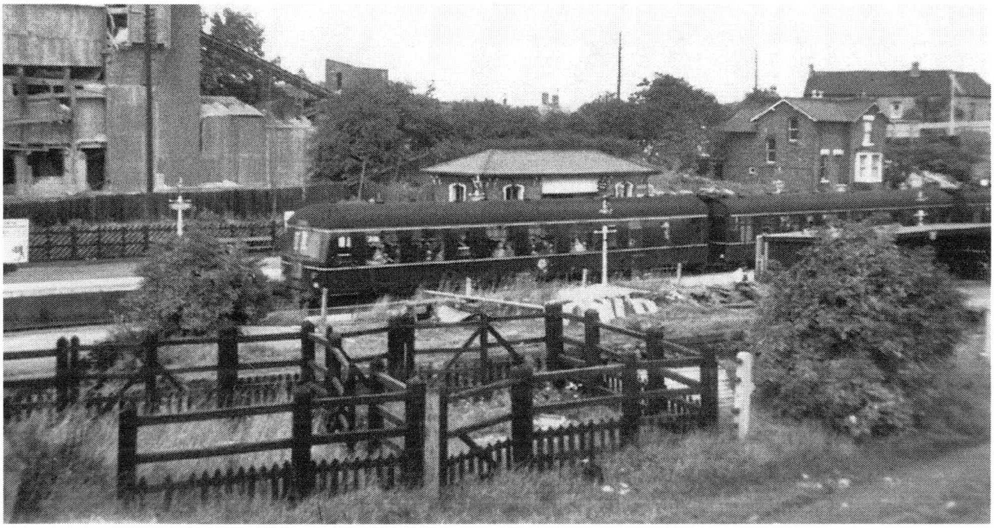

A view overlooking the cattle dock and platforms at Nuneaton Abbey Street station. Also visible is a section of the plant at Midland Quarry in the left background. The diesel train is heading towards Birmingham. You can see the waiting room on the Leicester platform and the stationmaster's house (which are extant today, and the stationmaster's house still occupied). (Maurice Billington)

housing development, which had started by 2010.

MOUNT PLEASANT COLLIERY, BEDWORTH

Operated in the 1830s and 1840s by William Thomas (who was also a lime-burner). This was connected with the Coventry–Nuneaton line in Bedworth by 1850. It was located behind the current Mount Pleasant public house, but closed in the early 1860s.

NATIONAL COAL BOARD LABORATORIES

Located in Corporation Street, Nuneaton. Prior to 1947, only three of the Warwickshire collieries had laboratory facilities. In 1947, after the National Coal Board was established, an area laboratory was set up at a disused mines rescue station at Wilnecote to look after the most southerly of the sixteen collieries in Warwickshire. In 1959, a new area laboratory was established in the former L. Tansey & Co. Ltd's needle works in Corporation Street, Nuneaton. Staff and equipment were moved from Wilnecote and the laboratory became fully operational on 27 July 1959. The scope of work carried out was as follows:

- Mine road dust analysis – fine particles of coal could cause an explosion.
- Mine air quality analysis to determine the level of contaminants in the underground air in the collieries.
- Study of airborne dust and the effect of dust of all types on the lungs of miners
- Coal analysis and quality checks on coal production.
- Water quality analysis involving testing source water used in boilers, steam-generating plants, water treatment plants and pithead bath discharges.
- Fuel and oil testing for machinery.
- Effluent testing

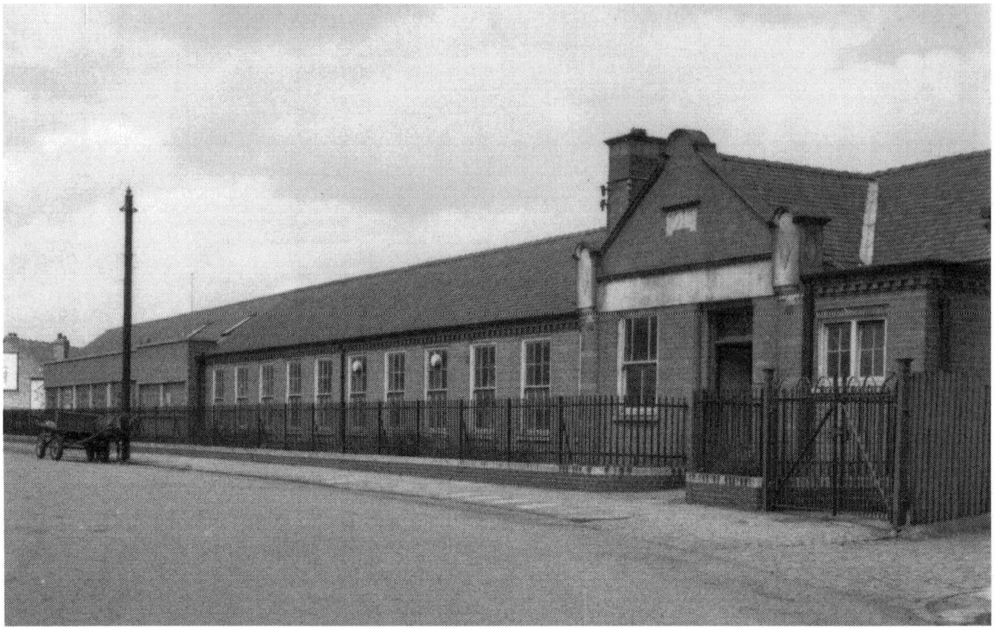

Tansey's former needle works in Corporation Street as it appeared in 1959, when the new NCB area laboratories were established there. Halford's store now occupies this site. (Author's collection)

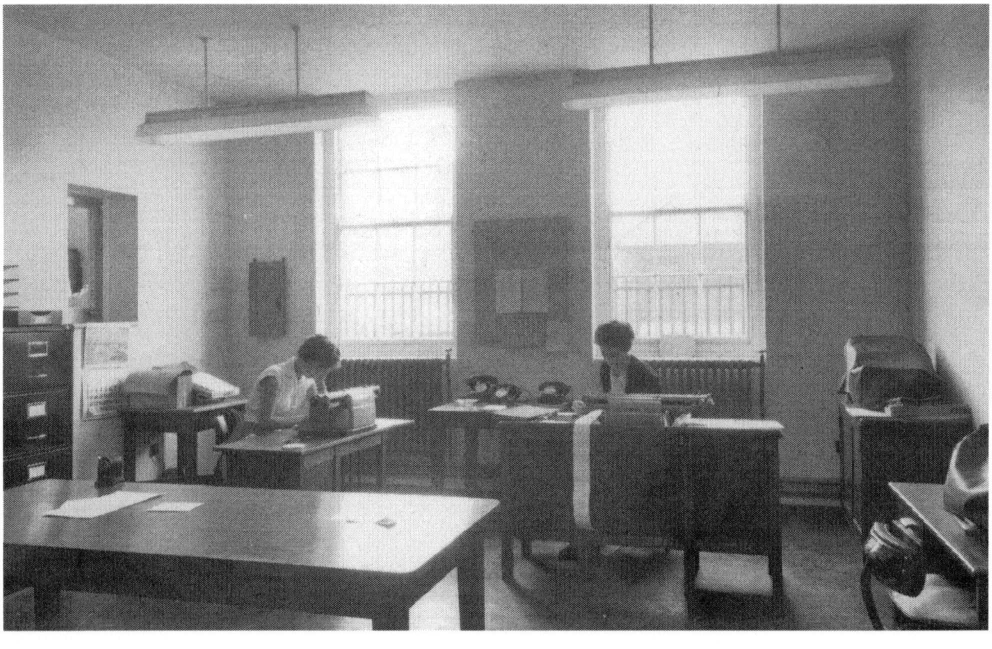

In the days before computers, offices like this looked sparse, but they still did the work just as well – albeit perhaps a bit more labour intensively than we would today. This is the area workshops office. (Author's collection)

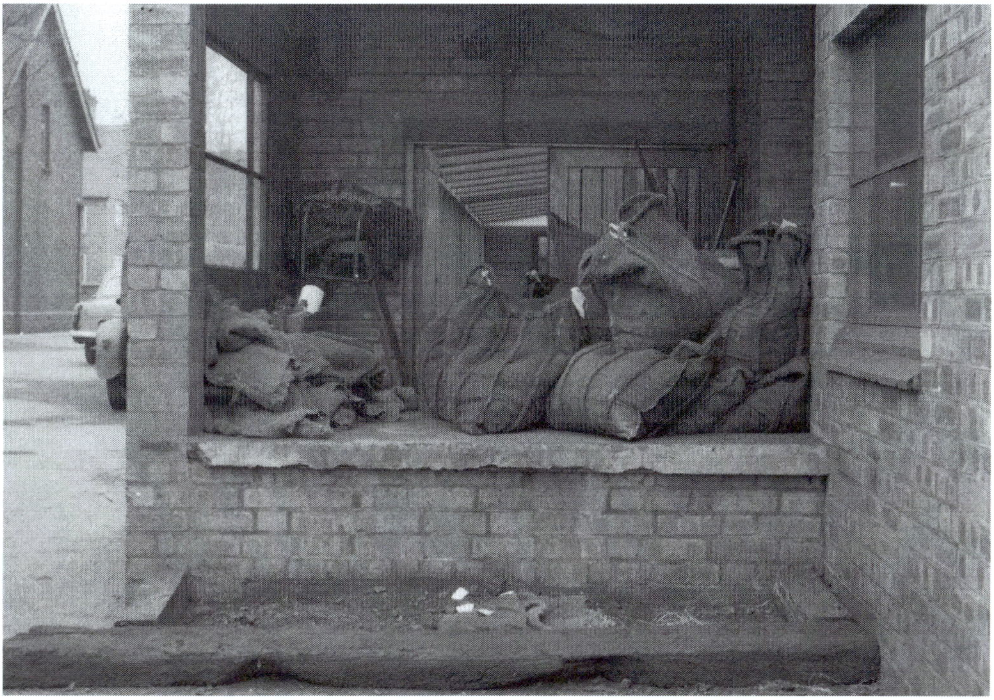

Coal was delivered from the various pits for testing in these rugged bags, which were heavily reinforced because of the rough handling they had to withstand. (Author's collection)

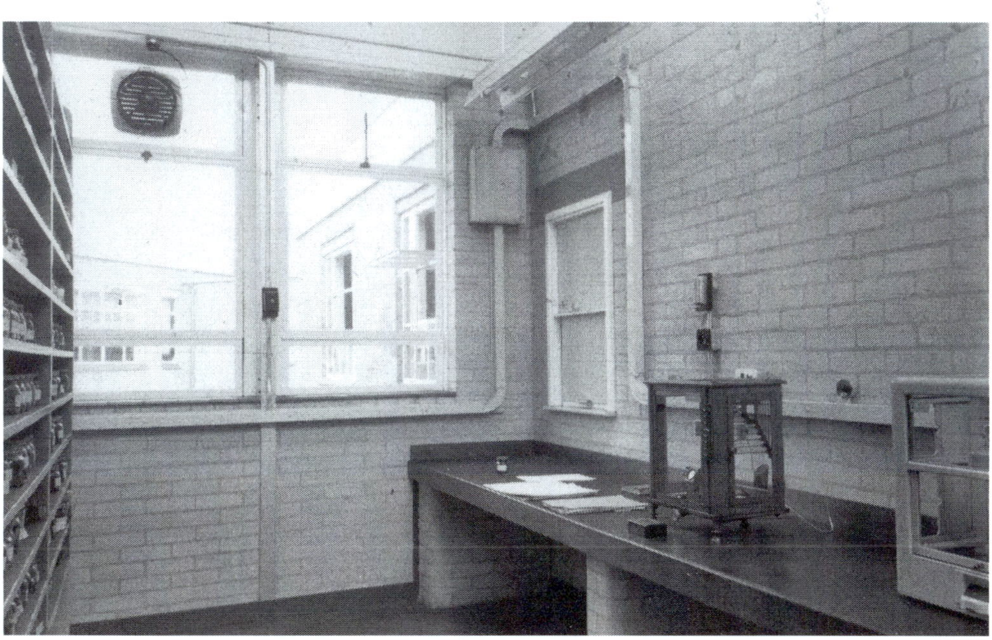

Interior of one of the testing laboratories. (Author's collection)

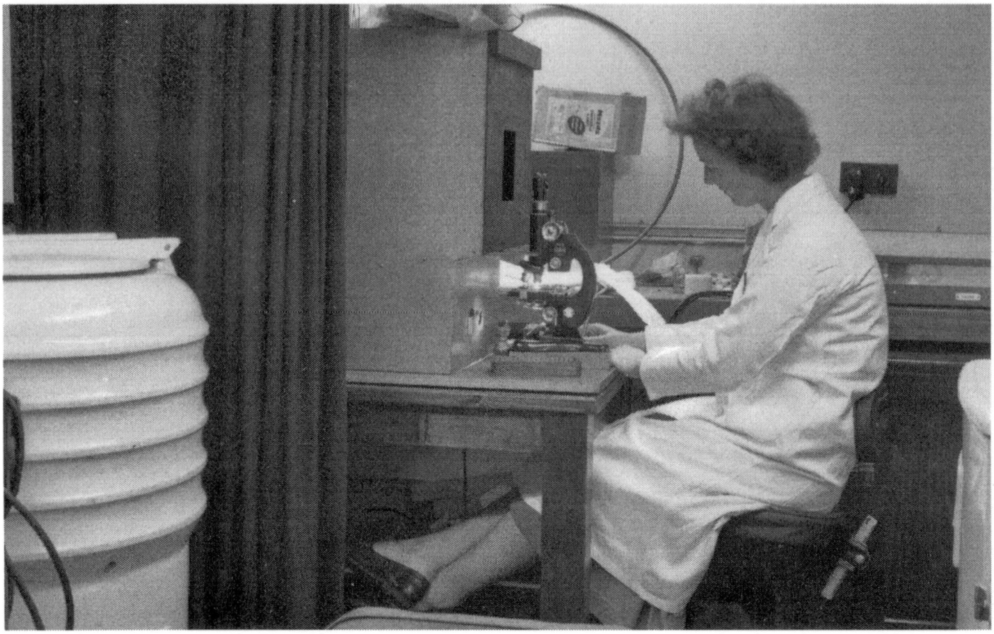

An attractive lady operates one of the devices used in the laboratory. It looks like a microscope. Staff were divided into different grades: Scientists (Grades I, II, and III), scientific technical officers (Grades I, II, III and IV), and Industrial and Clerical staff. (Author's collection)

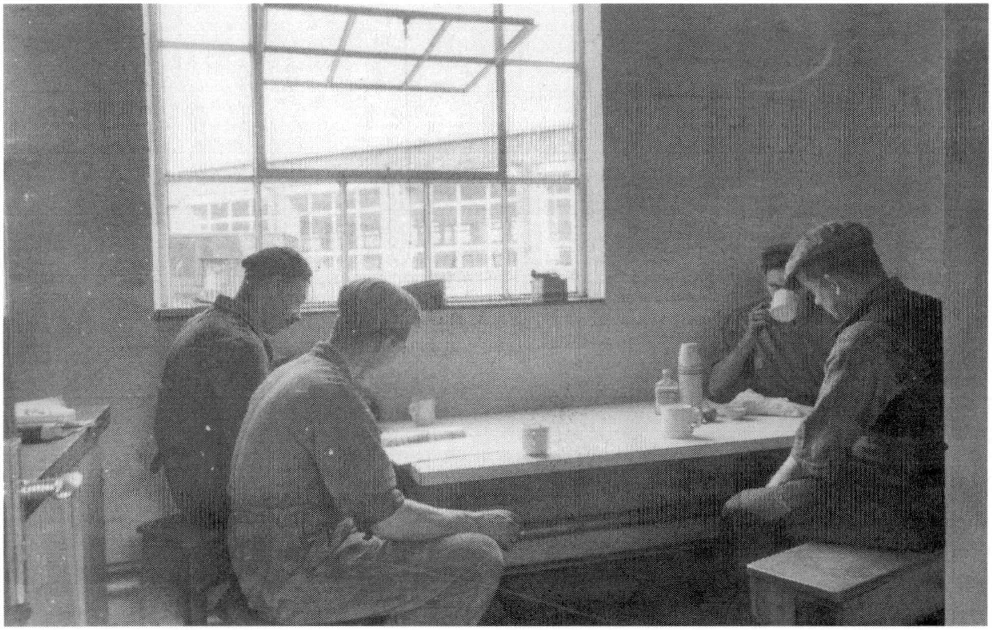

Sturdy workmen take a break in the laboratory canteens, presumably between humping coal bags around. (Author's collection)

- Purchasing and storage, particularly lubricants, fuels, paints and other consumables used in the local industry.

NEW COLLIERY, BEDWORTH

Owned by an E. Wootton and located near Charity Colliery in Bedworth, it was a latecomer among small independent collieries, being sunk in 1882 on the north side of the Charity Dock siding, and finally abandoned on 30 November 1894.

NEWDIGATE COLLIERY, BEDWORTH

Originally owned by F. A. Newdigate, this colliery was operating as Newdigate Colliery & Brickworks in 1900, then as Newdigate Colliery Co. Ltd until 1914, when it became simply Newdigate Colliery Ltd. It was operated by the National Coal Board from 1947 until its closure.

Sinking commenced in 1898 and coal winding started in 1901. In 1914, 1,174 men were employed below ground and 350 on the surface. In 1914, there was an underground fire which brought about a stoppage and the failure of the company. The company was re-formed in the same year. In 1938, there were a total of 1,249 men employed; in 1957, 1249 men. In 1960, the colliery employed 972 below ground and 404 on the surface, and on closure in 1982, just 235 men. A brickyard was also located next to Newdigate Colliery; this operated for some years.

A private siding arrangement was made with the LNWR for connection to the Coventry–Nuneaton line on 22 July 1897. By 1937, 1,500 tons per day was despatched, sent out in three fifty-wagon trains. In 1957, 374,600 tons of coal was produced. The colliery was working the Seven Feet, Nine Feet, Rider and Two Yard coal in 1960.

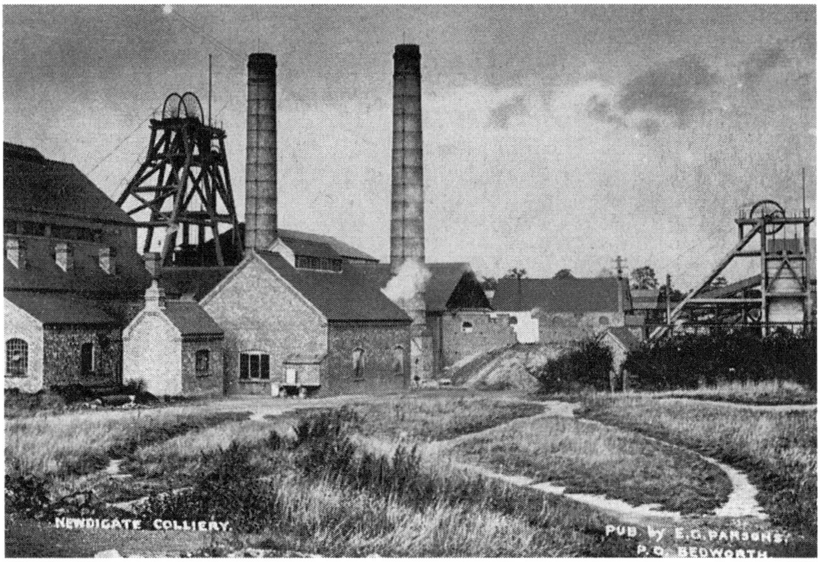

An early view, from an old postcard, of Newdigate Colliery, shortly after completion.

Next page, top: Newdigate Colliery mines rescue team kitted out ready for action. It is understood that the original picture came from Margaret Jacques, whose father-in-law, Bill Jacques, is the tall man second from left. He was a deputy at Newdigate. (June Robinson)

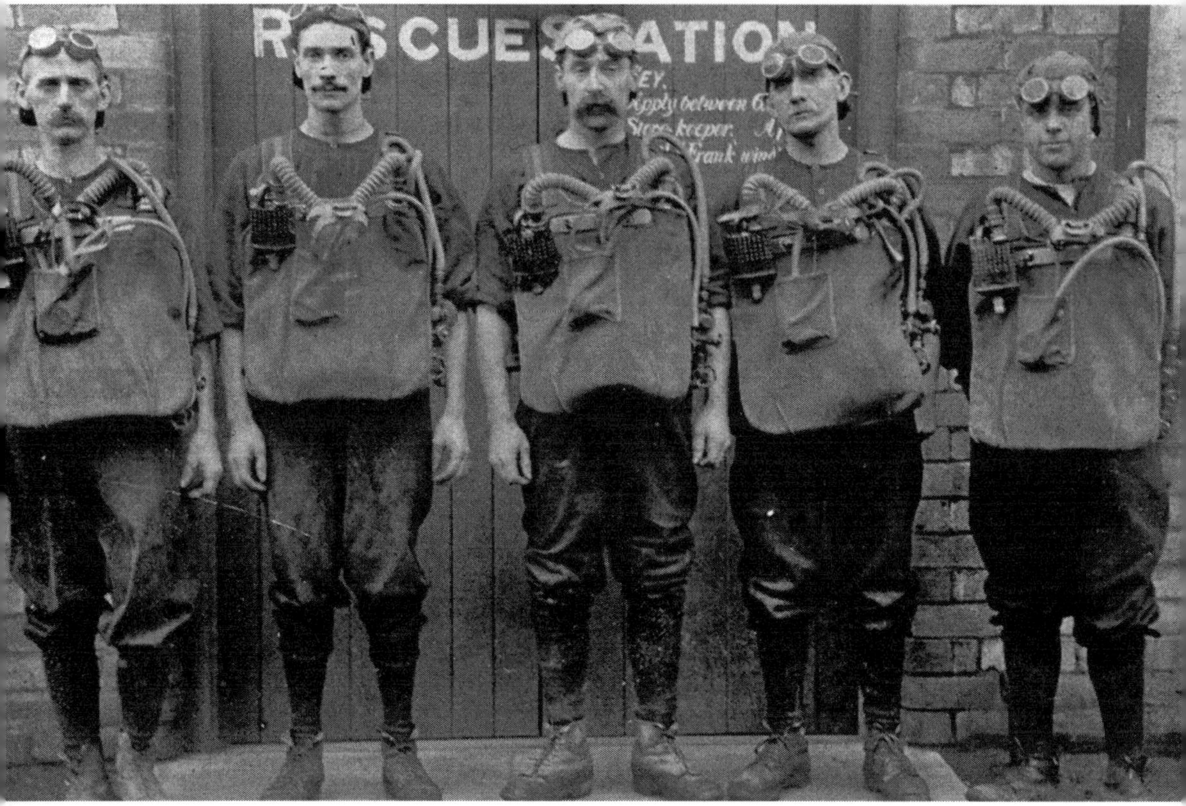

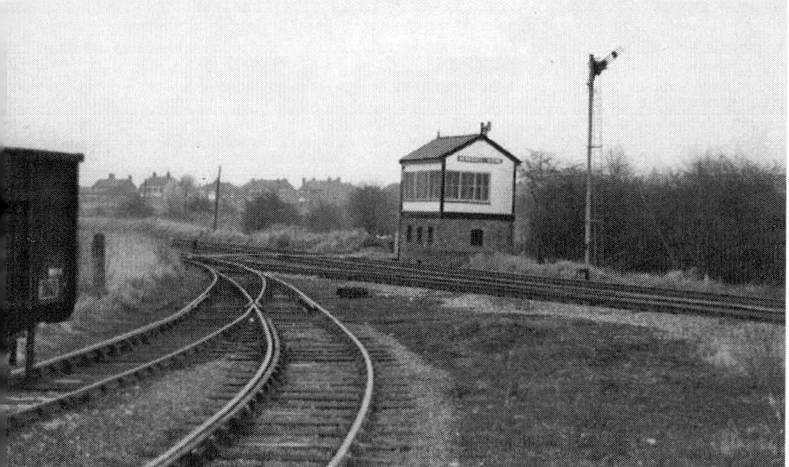

Newdigate sidings and sidings box. The track in the centre of the picture goes left towards Nuneaton, right to Coventry. The wagon stands on the Newdigate colliery branch and the photo was taken not long before rail traffic ceased. (Author's collection)

Opposite above: This is the same occasion, but the view was taken by the author with his back to the Coventry–Nuneaton line, not long before the cessation of rail traffic on 27 June 1981. We are looking here towards the colliery. (Author's collection)

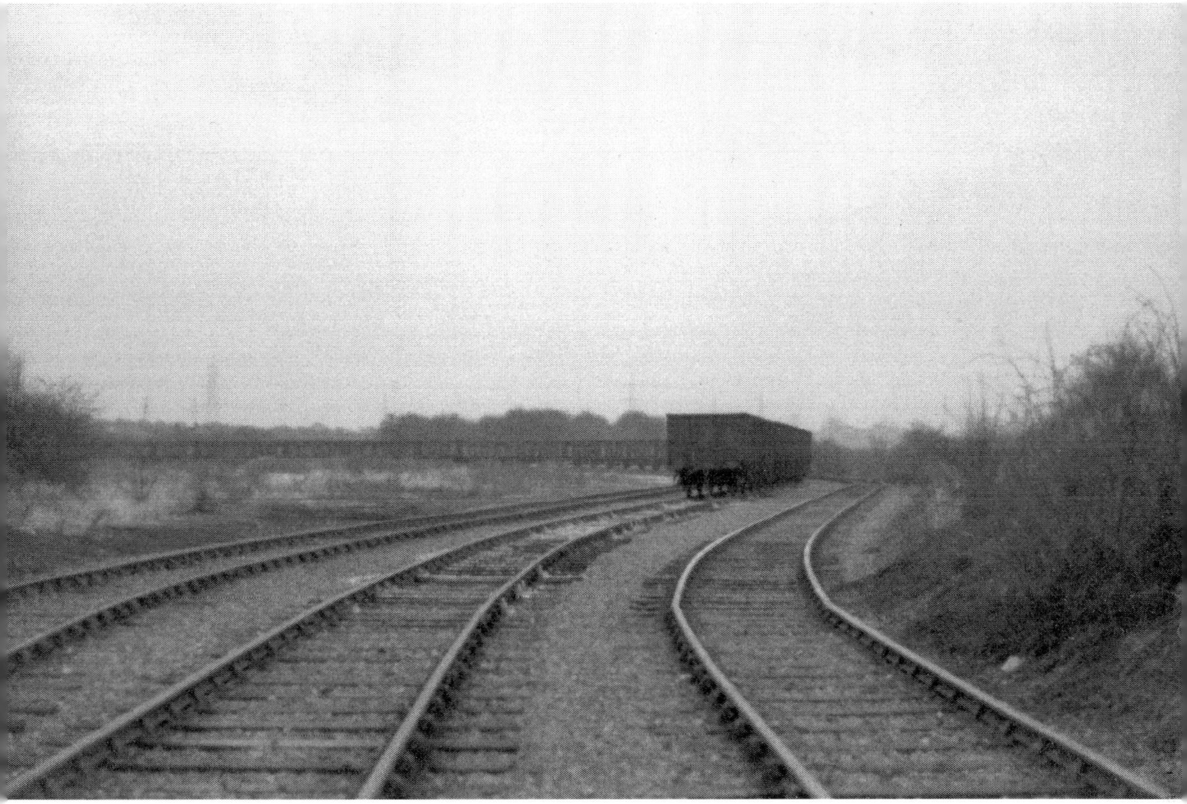

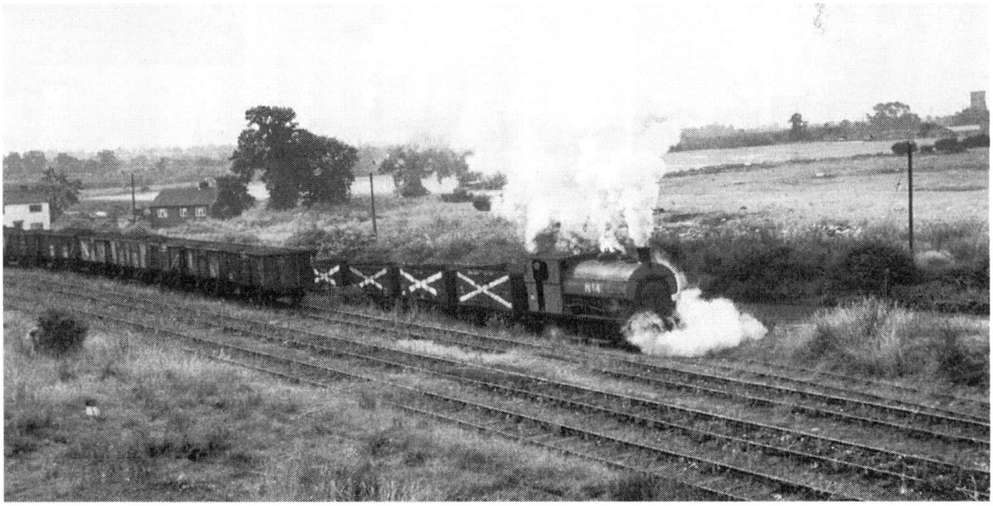

Peckett No. 4 (works No. 1787 of 1933) climbs from the canal arm having delivered a load of coal to be transferred into barges. It is a sharp gradient and No. 4 is working hard to lift the empty internal user wagons up the grade. The sidings in the foreground are the exchange sidings with the Coventry–Nuneaton line. The last boat to be loaded at this wharf was loaded in August 1966. No. 4 was out of use by January 1969 and scrapped around November 1971. (Mike Kinder)

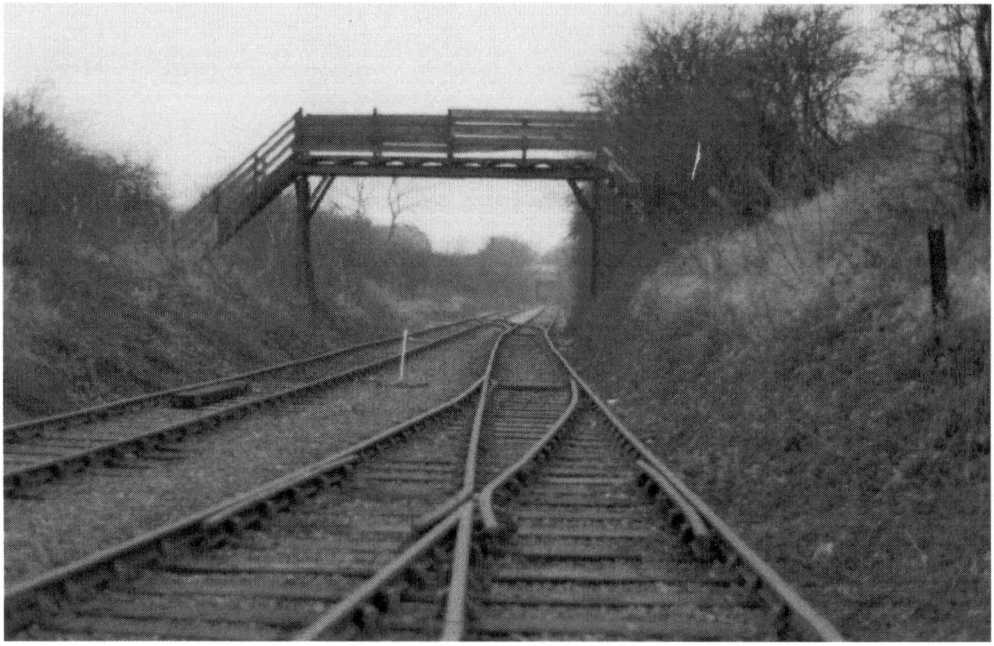

A view of the Newdigate exchange sidings, with a footbridge in the foreground that carries a footpath over the line. In the distance can be seen the bridge under the Coventry road in Bedworth, which is where the branch travelled up to the pit. The condition of this bridge led to the closure of the branch railway prior to the pit shutting. After the abandonment of the branch, all coal from Newdigate was sent underground to Coventry Colliery for transport. (Author's collection)

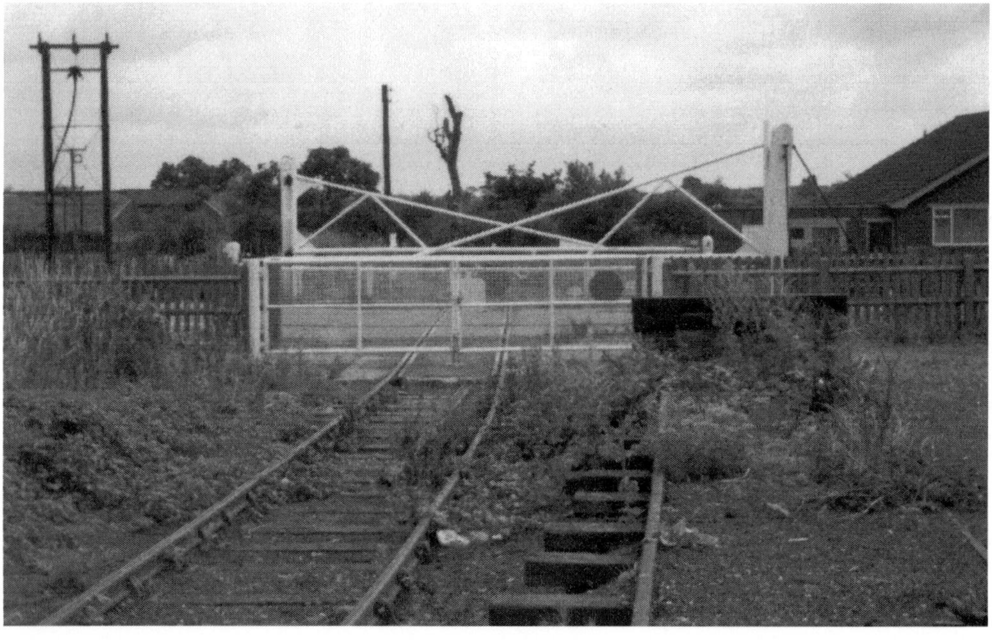

Opposite below: The level crossing where the Newdigate branch known as Smorrall Lane crossed. At one time, to the right of this view, Newdigate Colliery maintained a loco shed. A new shed was built at the colliery itself around 1961 and the old shed demolished. We are looking towards Black Bank and the exchange sidings with BR. The bridge under the Bedworth bypass road can be seen to the left of the gates. (Author's collection)

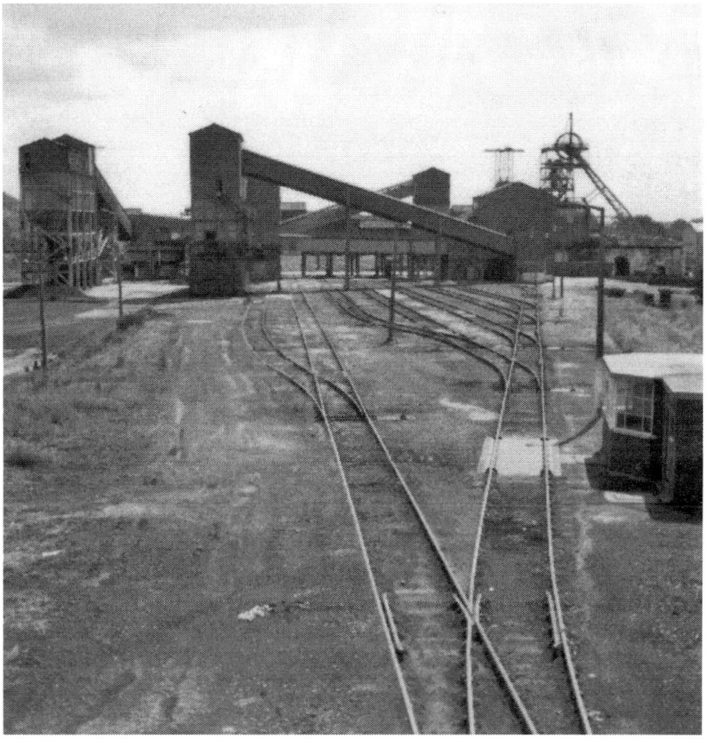

Right: The screens at Newdigate after closure. On the right is a wagon weighbridge. There was one of these at each end of the layout. (Author's collection)

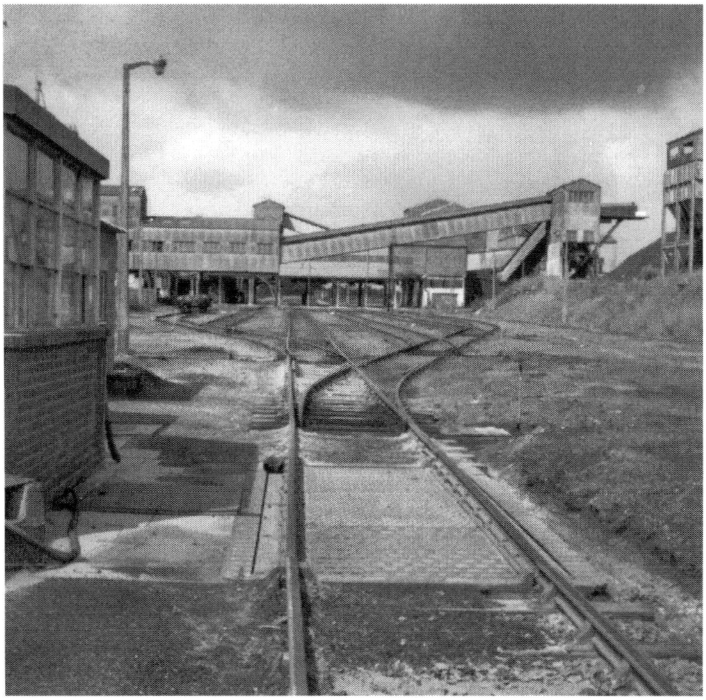

The colliery weighbridge, looking towards the screens. This was after the colliery's closure. (Author's collection)

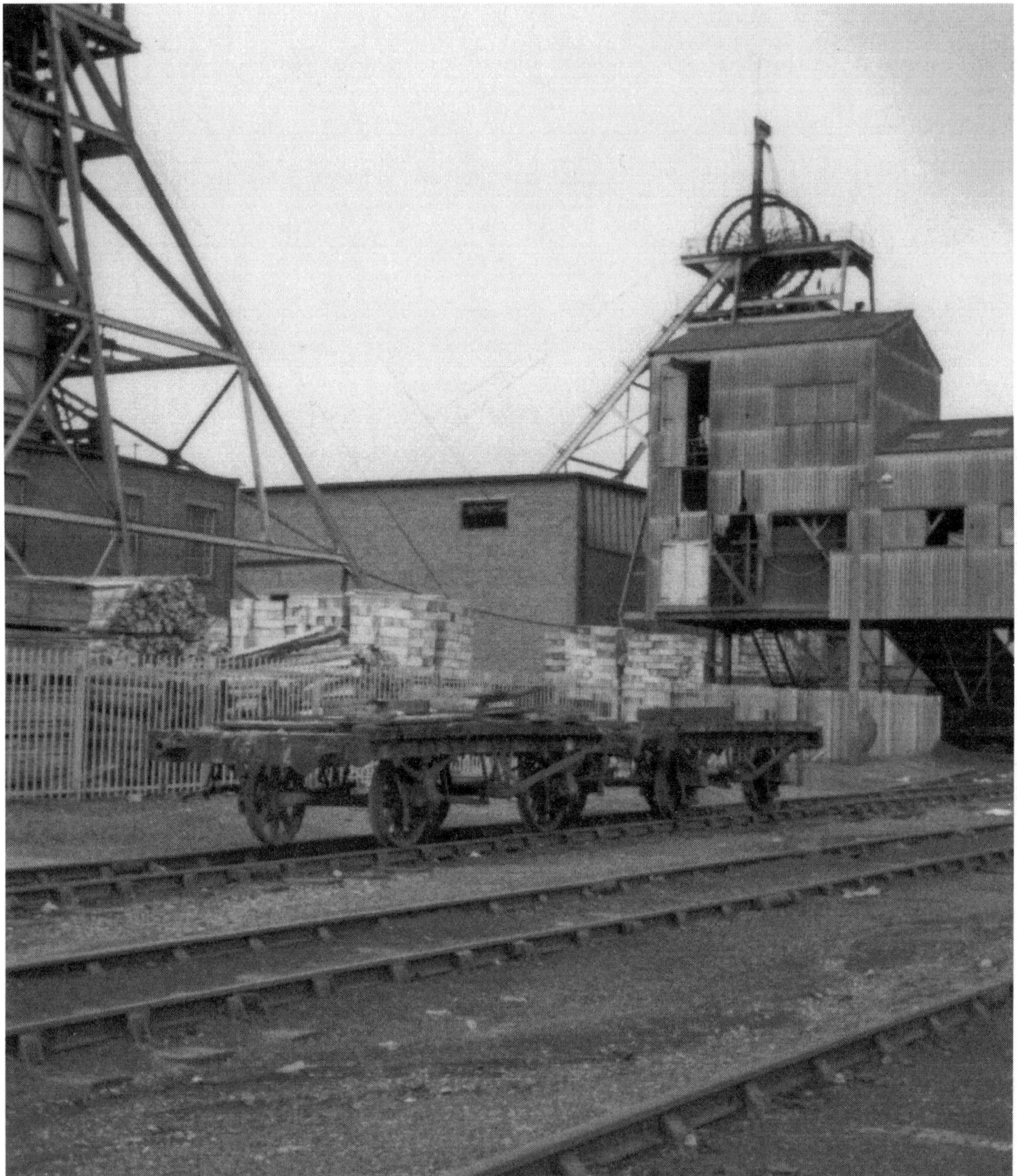

After the closure of the railway system, most of the rolling stock was removed, except for these two derelict wagons. They would be cut up on site by the demolition crew. They had apparently been ordinary mineral wagons with timber sides, but the wooden body had long been dismantled and they had been converted to internal-user flat wagons. From the look of their condition, they had not been in use for years. (Author's collection)

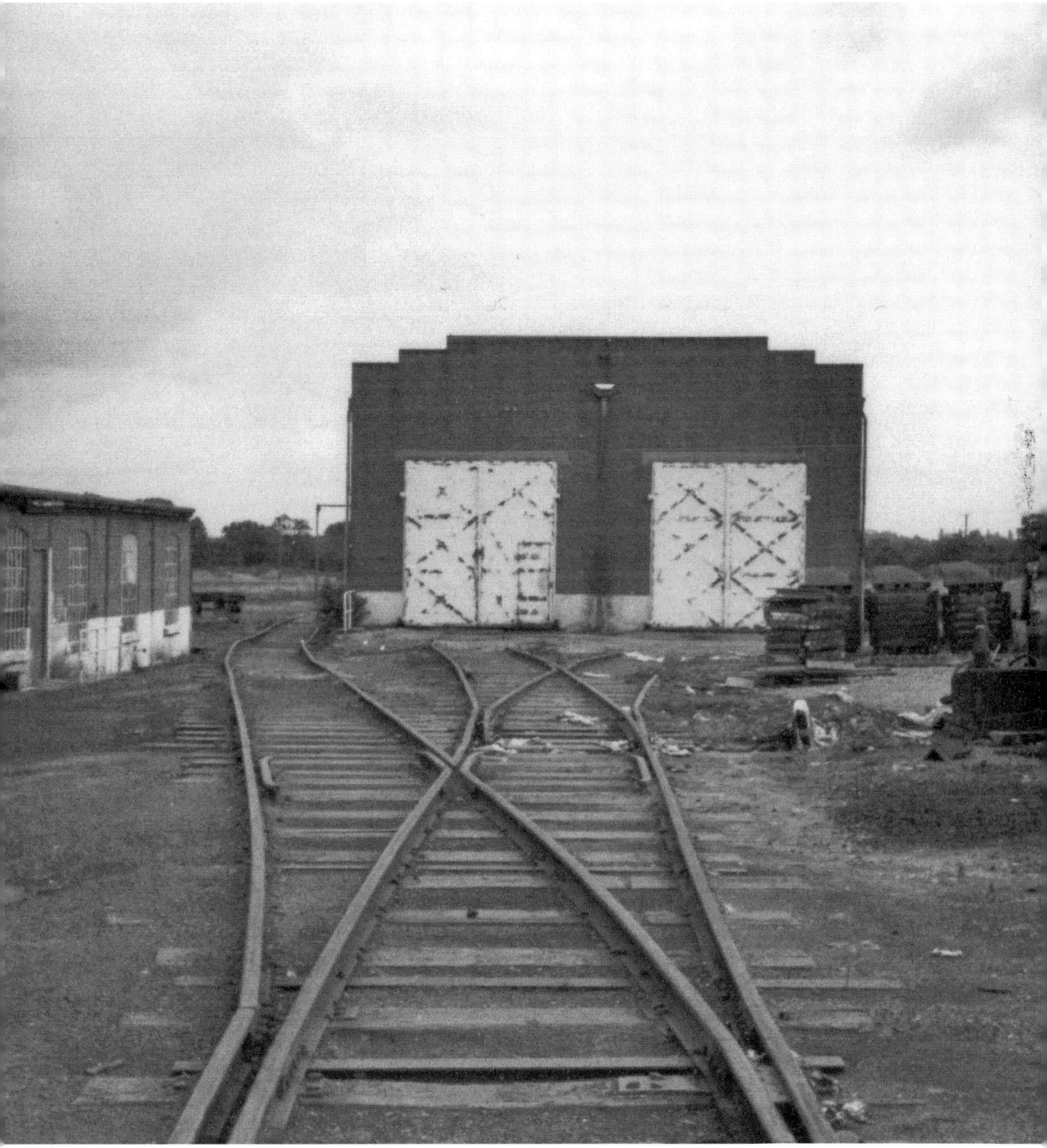

The colliery loco shed was a two-road affair and superseded the small single road shed around 1961, when a surface reorganisation of the colliery took place. The old loco shed formerly stood near the road crossing in Smorrall Lane. The alignment of the trackwork leaves a lot to be desired.

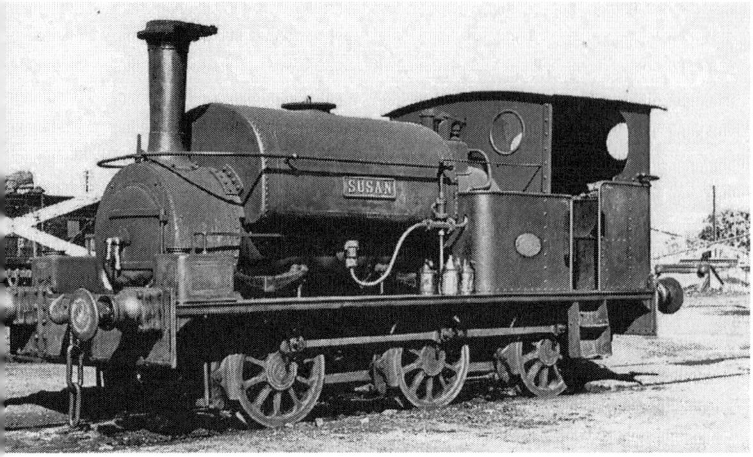

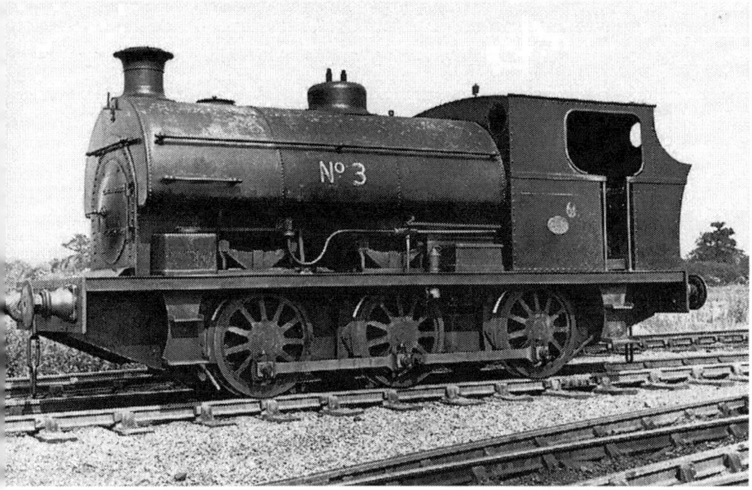

Top: Susan was one of the early locos at Newdigate. Built in 1902 by Hudswell Clark, their works No. 610 of 1902, it worked until it was worn out and scrapped on site in June 1958. (IRS)

Middle: Newdigate had two of these sturdy Pecketts. This is the first, which they acquired new in 1922 – works No. 1586. It put in many years of hard work before being cut up on site by Geo. Cohen Ltd in June 1968. (IRS)

Bottom: Coventry No. 1 made it into preservation, was last located at the Quainton Road preservation site, and is probably still there. It was a relatively modern engine, built by the North British Engine Co. Ltd in 1939 (their works No. 24564). *Coventry No. 1* was sent to Ansley Hall central workshops in 1963 and then went to Haunchwood Colliery in late 1965 or early 1966. It was returned to Newdigate in April 1967 and placed into the care of the Quainton Railway Society on 18 January 1971. (IRS)

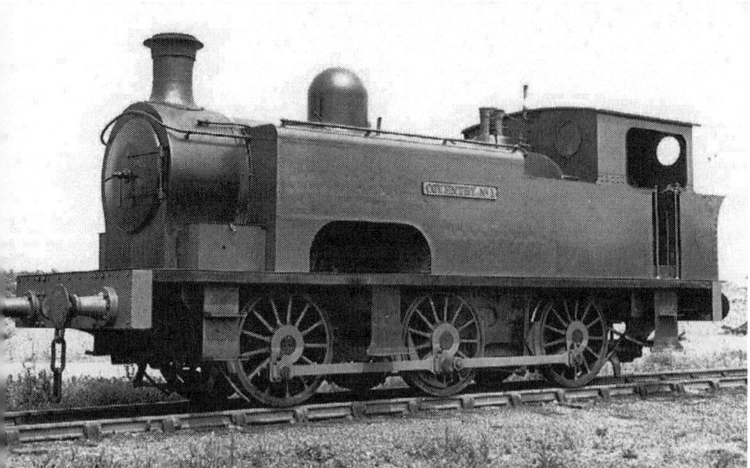

Opposite above: Coventry No. 1 in steam, together with one of the Pecketts. It may have been photographed here in the 1967–71 period after its return to the colliery from Haunchwood, although the Peckett being in steam would narrow this period down to 1967 to January 1969, when No. 4 was taken out of use. The narrow-gauge trackwork was 2 ft gauge. (Author's collection)

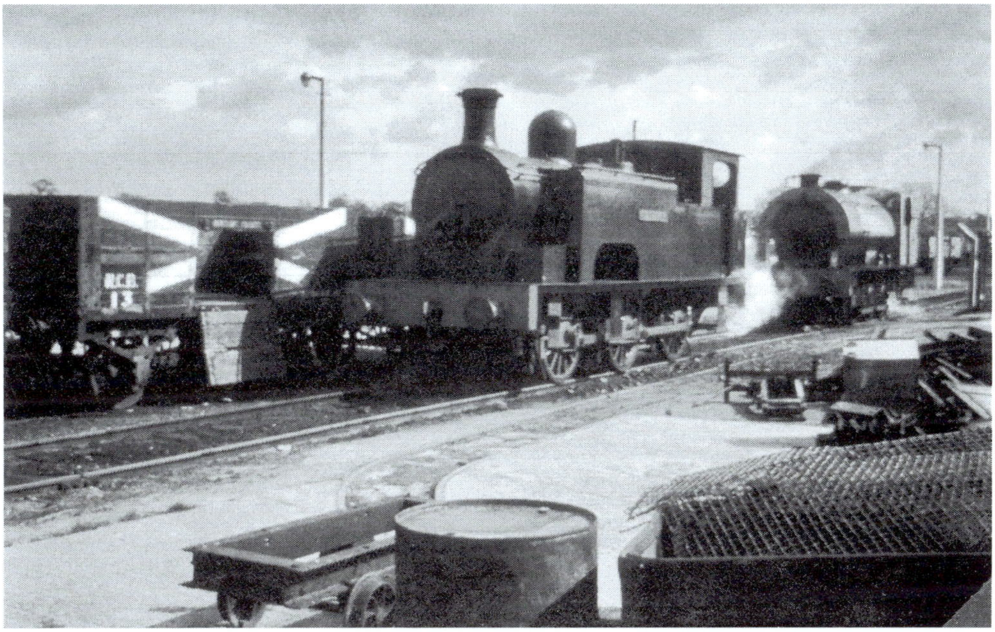

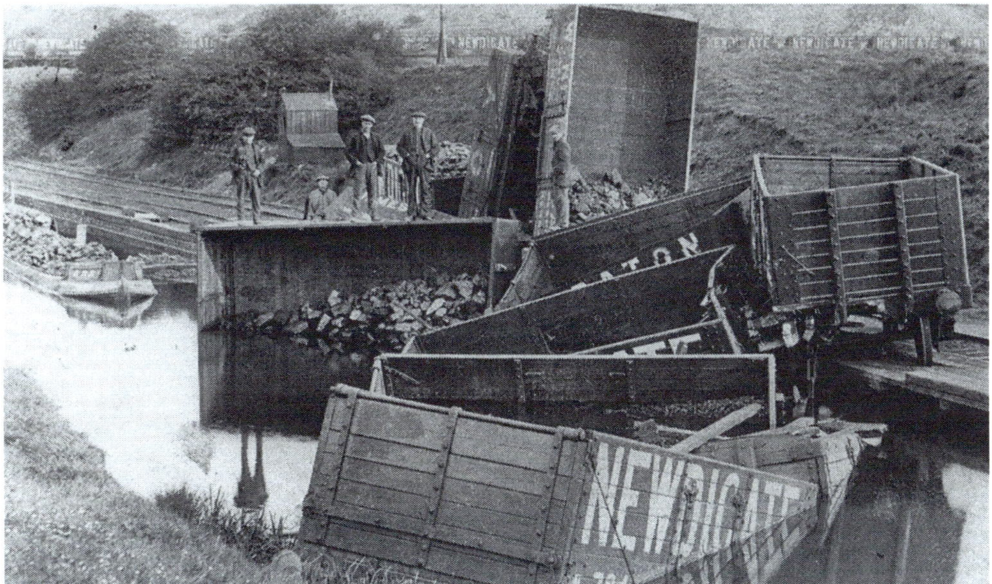

This is quite a well-known photograph, as for some reason this upset attracted the attention of a local photographer. What seems to have happened is that the raft of wagons has run away down the steeply graded incline to the canal wharf and fallen into the canal, discharging the loads of coal into the cut. It must have taken some time to sort the mess out. Those fully laden barges look to be trapped in the canal basin with their way out barred by submerged wagons. Just visible on the horizon is a line of wagons at the Coventry–Nuneaton line colliery exchange sidings. (Author's collection)

LOT 10

(PLAN NO. 2, TINGED PINK)

A VERY VALUABLE FREEHOLD PROPERTY,

SITUATE

ON THE ROAD FROM NUNEATON TO HARTSHILL,

COMPRISING

SEVERAL ENCLOSURES OF PRODUCTIVE

ARABLE, MEADOW, AND PASTURE LAND,

With the Coal and other Minerals, which are now being worked,

ALSO

ONE OF THE FINEST BLUE BRICK AND TILE YARDS IN THE COUNTY,

HAVING A DEEP DELF OF CLAY OF THE FINEST QUALITY,

TOGETHER WITH

BRICK-BUILT FOREMAN'S COTTAGE, FOUR DRYING SHEDS, ENGINE HOUSE,

SEVEN OVENS, OFFICE,

BLACKSMITH'S SHOP AND COW SHED,

TWO COAL SHAFTS, ABOUT THIRTY YARDS DEEP,

ALTOGETHER

CONTAINING TWENTY-SIX ACRES.

Let to Mr. Thomas Hardy, together with Lot 11, as Yearly Tenant, at

£64. 15s. PER ANNUM RENTAL.

And held at the very inadequate Royalty of 1s. 6d. per 1000 Bricks.

MORE PARTICULARLY DESCRIBED AS UNDER:—

No. on Plan.	Name of Field.	Cultivation.	Extent. A.	R.	P.
1	Brick and Tile Works, Old Clay Pit, and Waste Ground		3	2	8
2	Field adjoining Tile Works	Pasture	6	0	32
3	Cooper's Middle Close	Arable	2	1	30
4	Cooper's Bottom Close	Ditto	3	2	6
5	The Acres Close	Ditto	2	2	17
6	The Meadow	Meadow	2	0	35
7	Part of Pit Close	Pasture	0	2	23
8	Pit Close	Arable	1	3	3
9	Clay Pit		0	3	18
10	Field	Pasture	2	0	28
			26	0	0

It is seldom or ever that so compact and valuable an estate as Lot 10 comes into the market, offering a safe investment of capital with the guarantee of large and lucrative profits. It has been ascertained that there are Three Coal Seams, which have already been, or may immediately be, worked. The surface seam is now being sunken to supply the extensive Brick and Tile Works on the Estate. The Clay, which is of superior quality, produces the finest Blue Bricks, Tiles, &c., &c., that can be had in the Kingdom. The Estate commands an extensive frontage to a capital road, which leads to the Coventry Canal, at a distance of about 560 yards, and is within a Mile and a Half of the Railway Station, where daily communication can be obtained to and from all parts of the Country.

This is the sale notice for Nuneaton Colliery, dated 1864. The colliery had as part of its enterprise a brick and tile works that was typical of so many of these local collieries. In later years, Stanley Brothers would work the colliery and absorb the brick manufacturing side into their extensive business.

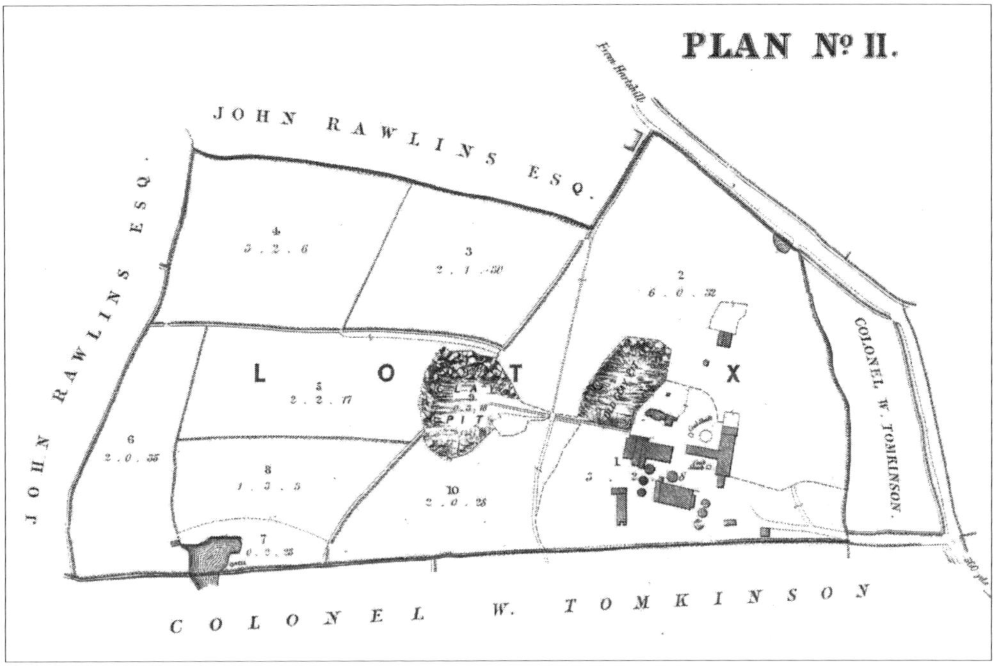

This plan of Nuneaton Old Colliery shows the layout of the site in the 1860s. The roadway to the right is now Haunchwood Road. Round brick kilns are evident as well as a former clay pit, close to the pithead buildings. It is thought that the brick yard was operated by the Williams family, who worked Haunchwood Colliery, and Charity Colliery. Later, offices were built fronting Haunchwood Road.

Among the major customers of Newdigate colliery had been Longford power station based in Coventry and John Dickinson, paper-makers, Hertfordshire. The colliery closed on 5 February 1982.

NUNEATON COUNCIL QUARRY

Nuneaton & Coton Urban District Council operated a small quarry at the top of Tuttle Hill for some years. At one time, it employed thirteen men. The manager was Mr F. C. Cook. The quarry has been abandoned for many years but has not been filled in as far as I can tell. There was a short tramway used to extract stone from the face to the perimeter. The stone was used for local road metalling and other municipal purposes.

NUNEATON OLD COLLIERY, NUNEATON COMMON

Owned by John Rawlins, landowner, the colliery was probably extant in 1864. Lessees included Thomas Hardy (by 1864), then Heath, Chadwick & Co. (known as 'Chadwick's Pit'), from around 1864 to 1870. Stanley Brothers worked the coal in the period 1877–1899. A Newcomen engine was erected here in 1736. Leases covering the period 1730–71 are still extant. The colliery closed in 1855 but reopened in 1868.

A tramway was built to connect the colliery with the Wash Lane Canal Wharf on the Coventry Canal. This was extant but disused in 1837.

The colliery had reopened by 1868, when there was an underground explosion there. In

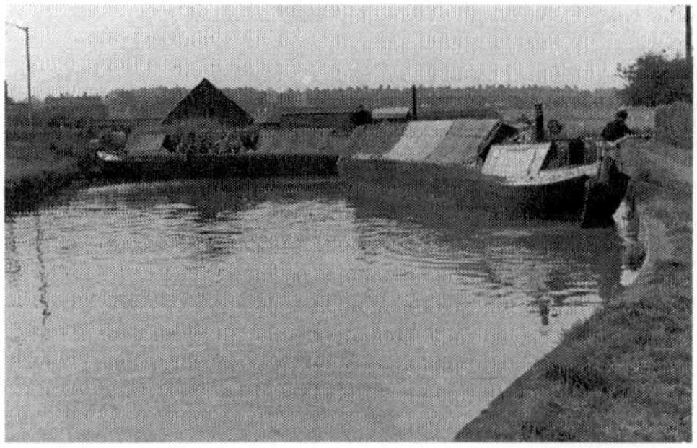

A wharf existed at the bottom of Haunchwood Road at the junction with Croft Road (formerly Swan Lane), which was the canal outlet for Nuneaton (Old) Colliery. This wharf was used until quite late, as can be seen here where a barge appears to be manoeuvring into position. The date is 17 September 1967. In an 1837 estate sale document, mention is made of a tramway that traversed what is now Haunchwood/Tompkinson Road down to this wharf from the colliery, although it seems to have been disused then. A lime kiln stood here too, and remains were still visible among the sheds arranged on the wharf in this view. (Geoff Edmands)

1896, Nuneaton (Old) Colliery employed 218 men underground and 61 on the surface; they worked the Slate and Rider coal-seams.

The colliery officially closed on 26 April 1899. The miners were then transferred to Nuneaton (New) Colliery at Whittleford.

NUNEATON (NEW) COLLIERY, WHITTLEFORD

Owned by Stanley Brothers Ltd (1900–22). In 1896, Nuneaton (New) Colliery was being sunk with twenty-four men underground and twenty-six on the surface. In 1910, it employed 1,115 men. In 1920, 1,392 were employed, but on closure in 1922 the figure had been reduced to 844.

After the colliery closed, the site was converted into Stanley's No. 7 brickyard, which operated between 1922 and 1940, when it too closed.

NUNEATON ENGINEERING WORKS

Based in Botterill Street and Tuttle Hill, Nuneaton. It was owned by Hall West & Co. and then Stanley Brothers Ltd. The works manufactured mining, brickyard, and stone-quarry equipment. It also manufactured the Stanley's coal-heading machine.

OLDBURY ENGINE PIT (POSSIBLY PART OF OLDBURY COLLIERY)

This was extant in the eighteenth century. No more is known.

OLDBURY COLLIERY

Owned by John Fletcher (no date), then John Barber (1734–1801), who worked the colliery until 1791; he was an inventor and took out several patents between 1766 and 1792. He is particularly credited with inventing the first gas turbine in 1791.

This colliery was probably one or more old-fashioned workings on land owned by the Dugdale, Ludford and Okeover families. The earliest date recorded is 1685. A Newcomen steam engine was erected in 1736. A tramway was laid down to the Coventry Canal and commenced operations in June 1817. It is not known when it was abandoned.

The colliery worked the Ell and Slate seams. I am not sure whether this should be grouped with Oldbury Spinney colliery, which worked the Doubles and Bench coal.

PARADISE LANE COLLIERY, STOCKINGFORD

Extant in 1881, according to a Census return. It might have been the same colliery as Pearce's Pit.

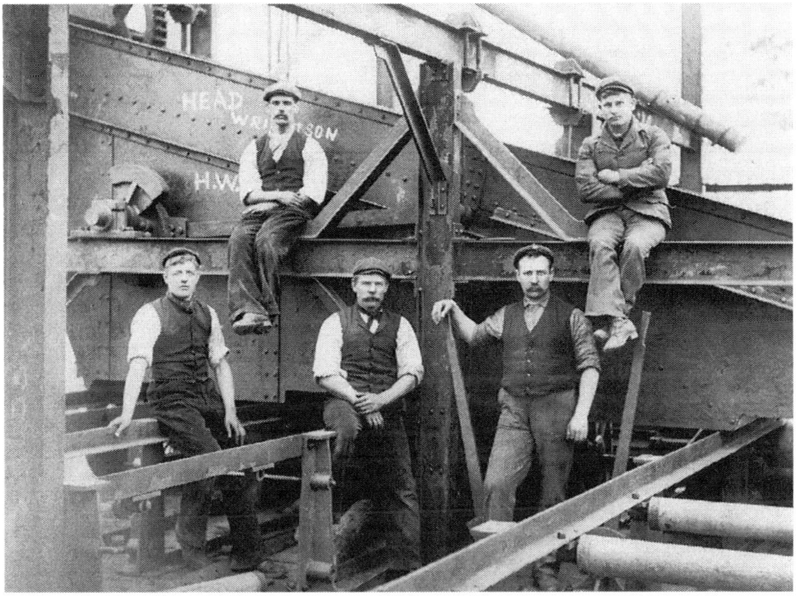

Sinking operations at Nuneaton New Colliery in 1899. Head Wrightson provided the headgear that is being erected here. The first shaft that was sunk filled with water and was converted into a well to provide drinking water for the local population of Nuneaton and district. (Author's collection)

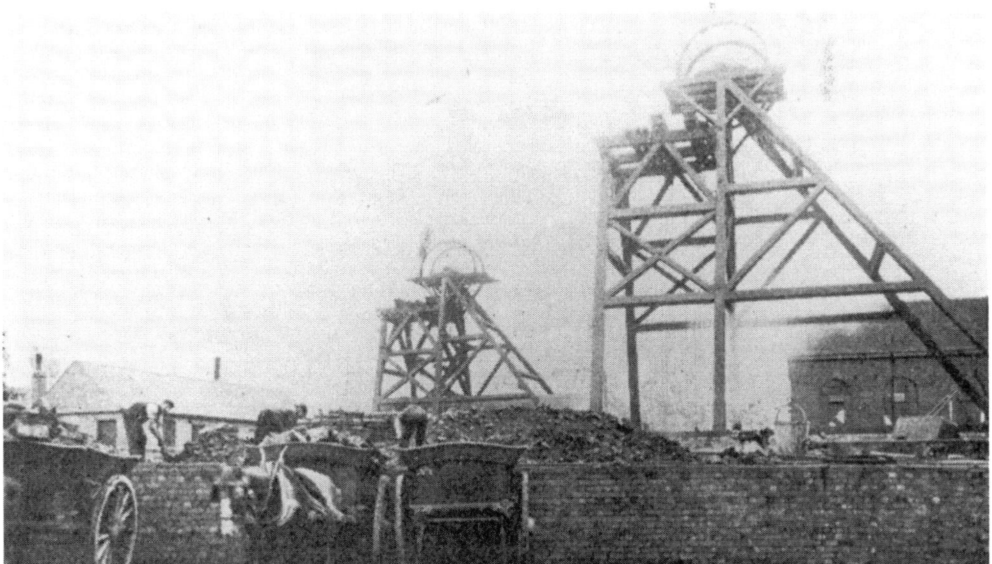

The land sale at Nuneaton (New) Colliery around 1900. This shows carts being filled for local distribution in the immediate area of Nuneaton. (John Nunn collection)

F. PARROTT

Paid rates on a brickyard in Bedworth in 1828. Location unknown.

J. PINKERTON

Paid rates on a brickyard, coal pits and stone pits in Bedworth in 1828. Location unknown.

JOSEPH PARKER

Parker was proprietor of the blue brick manufacturing works at Stockingford in the period 1866–1874.

EZRA PAUL

Extant in 1854. Builder, brick-maker, manufacturer of quarries, paving slabs, and drain pipes for sanitary purposes. Based in Attleborough (possibly at Attleborough stone quarry and brickworks). Samples of bricks with the name Paul incised in the frog are known to exist.

PEACOCK LANE BRICKYARD

Extant in 1845. Located off Peacock Lane (now King Edward Road), Nuneaton. The brickyard was situated in the middle of the trackbed of what is now the main railway line from Rugby to Stafford. The land belonged to William Stratford Dugdale but was rented to a local brick-maker, although we do not know which one of the Nuneaton brick-makers worked the yard at the time. It seems possible that a depression in Nuneaton cemetery, used for laying wreaths for people who have been cremated, may be a remnant of the clay excavations connected with this brickyard. The brickyard only appears on the track plan and section of the Trent Valley main line of the LNWR. It would have been largely obliterated by the construction of that line, commencing in 1845.

PEARCE'S PIT

Extant in the nineteenth century. Located near Nuneaton New Colliery, Stockingford. There is only a single map entry, so this may be a reworking of an old shaft.

PREMIER ARTIFICIAL STONE CO. LTD

Known as Moorwood, then later Globe Works, Chapel End. It then became Gibson & Preston's Patent Slab Works. This company was connected to the Stockingford branch railway line at Ireland & Knight's siding around 1897. A member of the Ireland family told me that their workings here were not successful, as the stone was of poor quality. That is why they moved to take over Mancetter Granite Quarries. It does seem, though, that the stone was of sufficient quality for crushing to make concrete. This was used by the Premier Artificial Stone Co. Ltd in their products. It is likely that production at the quarry here was short-lived and that stone came from other local quarries latterly. The spur from the branch also served Jee's Moorwood brickyard. The brickyard closed in 1912 (see separate entry). The spur seems to have been abandoned in 1915, although not lifted until the 1930s. There was a narrow-gauge, rope-worked tramway connecting the railway loading dock, the works and the quarry face. Premier Artificial Stone Co. Ltd made paving slabs, kerbs, house lintels, and other concrete building products. The concrete company moved to the site of Stockingford Colliery in the 1930s, taking over the old colliery site and continuing production there until the 1990s.

JOHN ROBINSON (1822–1896)

A John Robinson (probably the father) operated a brickyard on the Common, Stockingford, by 1828. He resided in Church

Street. A brickyard was still operated by a John Robinson (presumably the son of the original John) as late as 1874.

ROWLAND ROBINSON

Listed in 1850 as a brick-maker and farmer in Church Street, Nuneaton.

THOMAS ROBINSON

Opened a blue brick and tile yard in Swan Lane and also sunk an 80-foot-deep coal shaft.

GEORGE SIDWELL

Brick-maker in Chilvers Coton around 1818. George is said to have emigrated to America, where he sold his brick-making equipment patent.

ROBERT SIDWELL

Listed in 1850 as a brick-maker in Heath End, Chilvers Coton.

STANLEY BROTHERS LTD

Formerly owned by David Wheway, the company was offered for sale in January 1869 after his death in 1868. The company traded as Stanley & Broadbent in 1869, then as Stanley Brothers (1869–1895). Stanley Brothers Ltd (1895–1988) became part of the Higgs & Hill Group in July 1961. The remaining business was sold to the rival Leicestershire terracotta manufacturer Redbank Manufacturing Co. Ltd in 1986. All sites except No. 5 yard have been redeveloped for housing.

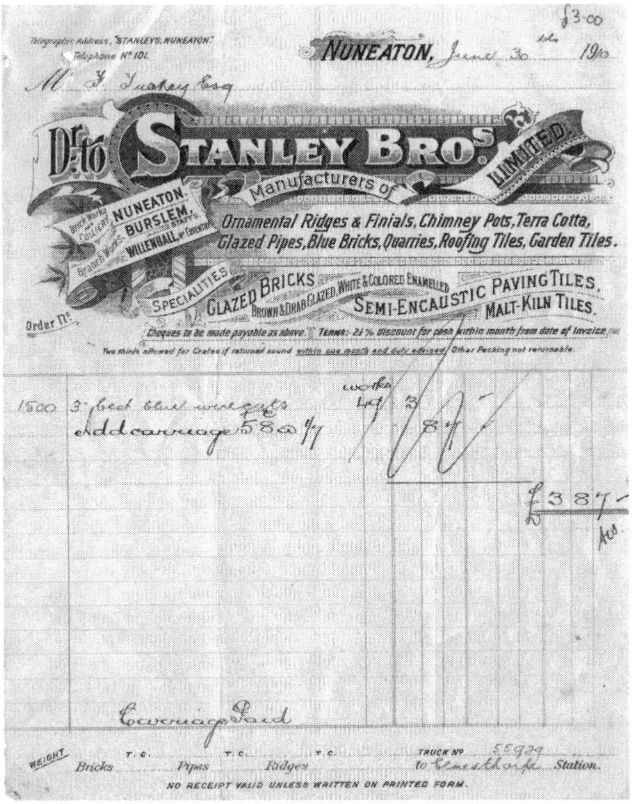

Telegraphic Address, "STANLEYS, NUNEATON."
Telephone No 101.

NUNEATON, June 30th 1910

Dr. to STANLEY BROS. LIMITED.

Manufacturers of

Brick Works and Colliery NUNEATON. Branch Works BURSLEM, STAFFS. WILLENHALL, nr COVENTRY.

Ornamental Ridges & Finials, Chimney Pots, Terra Cotta, Glazed Pipes, Blue Bricks, Quarries, Roofing Tiles, Garden Tiles.

Specialities: GLAZED BRICKS, Brown & Drab Glazed, White & Colored Enamelled. SEMI-ENCAUSTIC PAVING TILES, MALT-KILN TILES.

Order No.

Cheques to be made payable as above. Terms: 2½% discount for cash within month from date of invoice.

Two thirds allowed for Crates if returned sound within one month and duly advised. Other Packing not returnable.

1500 3" best blue wirecuts ... 49/3
Add carriage 58 @ 1/7 ... 8 7

£3 8 7

Carriage Paid

WEIGHT: Bricks T. C. Pipes T. C. Ridges T. C. TRUCK No 55929 to ... Station.

NO RECEIPT VALID UNLESS WRITTEN ON PRINTED FORM.

In 1895, a limited liability company was founded with branch works at Burslem, Stoke-on-Trent, Willenhall near Coventry, as well as Stockingford. A colliery was founded at Bedworth.

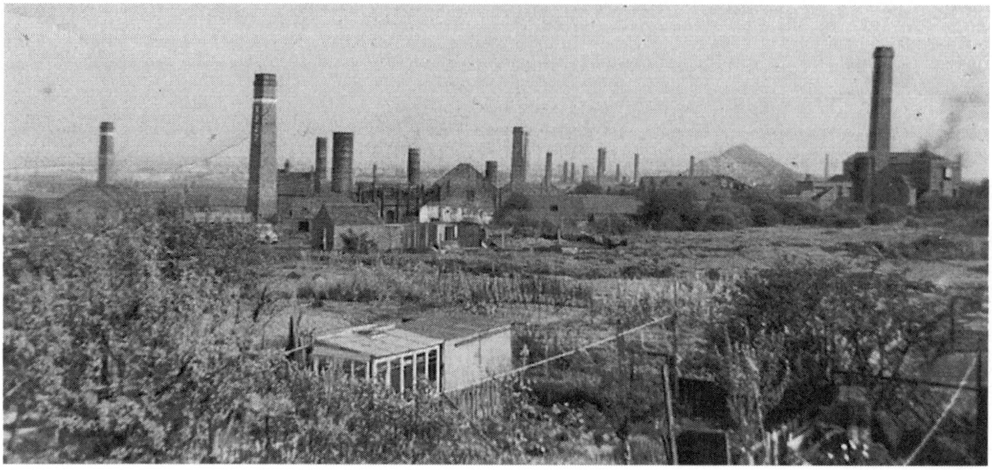

This is an excellent photograph taken by Vic Holloway from his rear bedroom window in Haunchwood Road, looking over towards Stanley's Nos 1 and 2 yards. The sea of chimneys beyond are Stanley's Nos 3 and 4 yards over towards the Heath End Road. The tip on the horizon is Griff No. 4 Colliery's slag heap. The nearest chimney has the date 1937 in the brickwork set in glazed bricks, indicating the year it was built. To its right are the distinctive features of a German Hoffman kiln installed in 1937, but it ceased to be used during the Second World War due to lack of spare parts available from Germany. In 1895, Stanley Brothers consumed 60,000 tons of coal annually, most of which came from their own production. In 1909, 3,000 people were employed by the company. (Vic Holloway)

In 1857, the yards later operated were in the hands of four parties who made blue bricks, glazed pipes and roofing tiles. One yard was owned by Handley & Wheway. On the death of David Wheway in 1868, his executors sold the brickyard to two Leicester businessmen trading as Stanley & Broadbent. In Nos 1, 2, and 3 brickyards was a fireclay pit employing six men underground and two on the surface. Stanley Brothers' brickyard estate covered 145 acres by 1900. The driving force behind Stanley Brothers was Reginald Stanley (1838–1913). He was borne in Hayle in Cornwall and migrated to the United States of America in 1857 to seek his fortune. After an epic decade in America, which included a period of great turbulence during the American Civil War, he went gold prospecting out west. While out there, he became one of the founding fathers of Helena, the state capital of Montana. On his return to England in 1869, he found his brother Jacob and cousin in partnership. Reginald Broadbent was in failing health, so Reginald Stanley bought into the partnership and in 1872 his brother relinquished his day-to-day involvement, but still retained a financial interest, as indeed did the Broadbent family. They were shareholders and sleeping partners for many years. The Broadbents owned a builder's merchant's business in the city of Leicester. When Stanley Brothers became a limited company in 1895, it was said to have 4,000 customers. Its yards and products were as follows:

- No. 1 Yard – red ridge tiles and chimney pots
- No. 2 Yard – salt-glazed stoneware pipes
- No. 3 Yard – multiplex quarries, super paving tiles, common red bricks
- No. 4 Yard – glazed bricks and sinks, malt kiln tiles
- No. 5 Yard – blue bricks and paviors
- Nos 6 & 7 Yards – common bricks

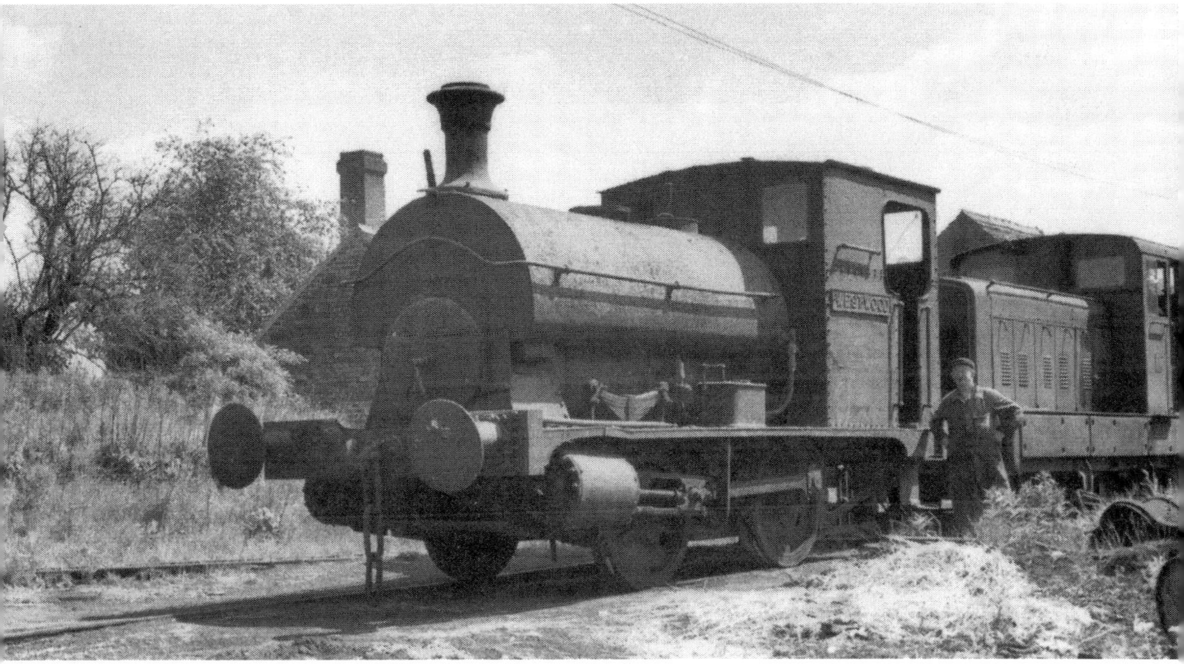

Stanley's purchased a Yorkshire Engine, No. 257 of 1875, second-hand from dealers and repairers James Tait Jnr & Partners in 1920. It had worked earlier at the Frodingham Iron & Steel Co. Ltd and was named after the then company secretary – William Westwood. Some years after it was last used, *Westwood* was pulled out of its shed for inspection by a visiting rail enthusiast, with the 'new' replacement diesel shunter. The four-wheel diesel mechanical by Ruston Hornsby was purchased new in 1950 to replace *Westwood*, which was laid aside shortly afterwards. *Westwood* was sold to G. W. Bungey Ltd in November 1961 and then resold to the scrap merchant George Cohen, Sons & Co. Ltd of Kingsbury and cut up in 1962. The diesel was sold at the same time to G. W. Bungey and sold on to Sheepsbridge Stokes Ltd of Chesterfield in Derbyshire.

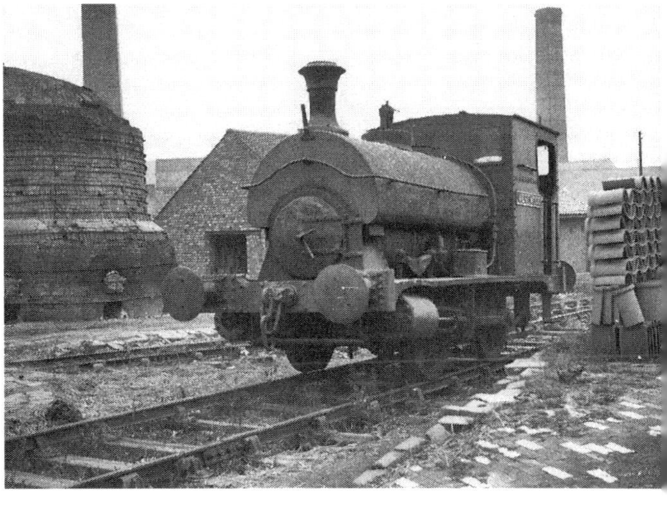

Westwood is placed into position, ready for cutting up just a few days before being scrapped in 1962. This photo has been included to illustrate the familiar environment of the brickyard, with stacks of ridge tiles to the right and one of the common beehive kilns to the left. (Geoff Edmands)

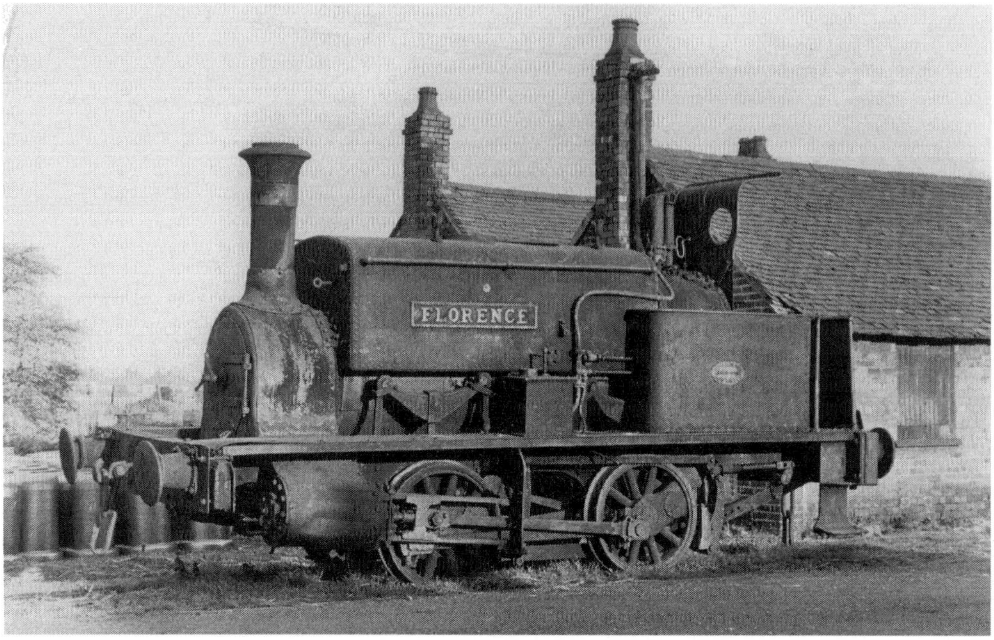

Stanley Brothers loco *Florence* was withdrawn immediately after the Second World War. Built by Manning Wardle, works No. 593 of 1877, it was originally owned by the Premier Cement Co. Ltd of Irthlingborough in Northamptonshire. This photograph was taken when it was laid aside, thoroughly worn out, not long before scrapping in 1950. It had been used up until the Second World War on general duties around the works. Its short wheelbase was ideal for the sharp curves and light trackwork prevalent in the various yards. During the last war *Florence*'s main duties were transferring wagons from the Griff branch sidings into the sidings at the Nos 3 and 4 yards, where unused kilns and sheds were being used to store food and vehicle parts for the Ministry of Food and the Daimler Company. The empty buildings had been commandeered by the government as buffer storage, away from the bombing threat over the big cities such as Coventry. Both works did not reopen for production after the war and were demolished in 1963/64. (K. J. Cooper, IRS)

Production finally ceased on 31 January 1988. Its commercial records were acquired by the Nuneaton Civic Society and are on permanent loan to Warwickshire County Records Office.

STOCKINGFORD COLLIERY

Operated as Stockingford Colliery Co. Ltd from 30 December 1868. Production commenced in 1872. Generally, it had a fairly uneconomic history and exchanged hands several times. In 1896, the colliery employed 141 people underground and 67 on the surface. It worked the Rider coal. In 1910, it employed 393 men. In 1920, 499 were employed. By closure on 16 March 1928, 460 were employed. A brickworks was also in use and operated up until 1894, but was disused thereafter.

SWAN LANE COLLIERY

Owned by Cooper & Handley, then Handley & Wheway. David Wheway (?–1868) was the son-in-law of Walter Handley (1783–1864). The colliery was operated by Stanley Brothers in the period 1872–78.

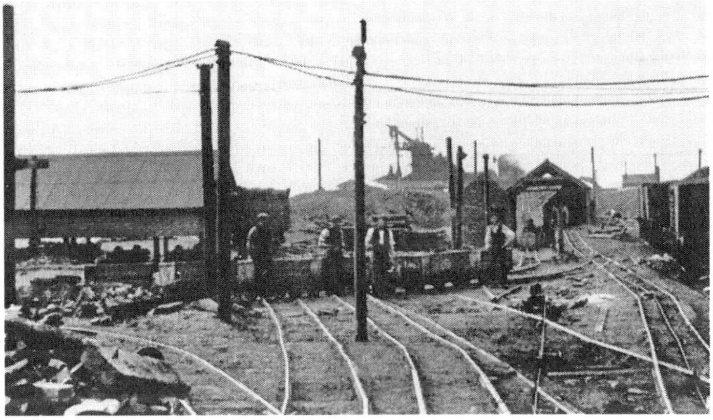

The pithead at Stockingford Colliery (familiarly known as 'Drybread', which may be derived from the Coal Mines Regulations Act of 1872, known as the 'Drybread Act', which came into force at the time of sinking). It was a drift mine, as the shafts went down at 45 degrees. It was the only drift mine in Warwickshire at the time. The pit yard had the appearance of a junkyard, with baulks of timber, old rail and general detritus scattered about. The state of the trackwork indicates re-railing was a regular preoccupation with surface workers. There are suggestions in recent years that Stockingford Colliery had steam engines for shunting purposes, although I have no means of confirming this theory. (The Stephenson Locomotive Society have produced a document that states that two locomotives were to be included in the closure sale of plant and equipment in October 1928.) The former general opinion was that it was always shunted internally by about a dozen Clydesdale horses. The MR worked the assembled wagons away from the exchange sidings down the branch to be sent out over the main line from Stockingford yard. For several years after closure, several derelict Stockingford Colliery wagons lay in the exchange sidings. (Author's collection)

Swan Lane Colliery was part of the business of Stockingford farmer Walter Handley, who also owned a brickyard and the Black Swan pub, which he opened in 1830 – hence the name given to the stretch of roadway between the Coventry Canal and the location of the pub – Swan Lane, known today as Croft Road. The executors of David Wheway sold the brickyard to the partnership of Broadbent & Stanley, later to become Stanley Brothers Ltd. The colliery was worked out by 1878. Walter Handley is known to have had a brickyard in Wilnecote, near Tamworth, in 1825.

ROBERT TYLER

Tyler was recorded in 1564 as being a roofing tile manufacturer in Nuneaton town.

VICTORIA COLLIERY

Owned by Thomas Ball Troughton & Henry Lea (in partnership), then Thomas Marshall, John Hartopp and John Oliver (in partnership) by 1865. Additional shafts in the immediate area were:

- Richards Pit
- Biggs Pit
- Deep Pit
- Bassett Pit
- Harrison's Pit
- Coggs Pit

Victoria Colliery was on the site of Victoria Farm, on land owned by a family called Hopkins. The site of the colliery today lies at the rear of the Greyhound pub at Sutton Stop, Hawkesbury, and at the rear of a neat set of cottages with distinctive chequer-pattern brickwork – these are former collier's cottages.

The colliery employed the famous nineteenth-century workers' champion

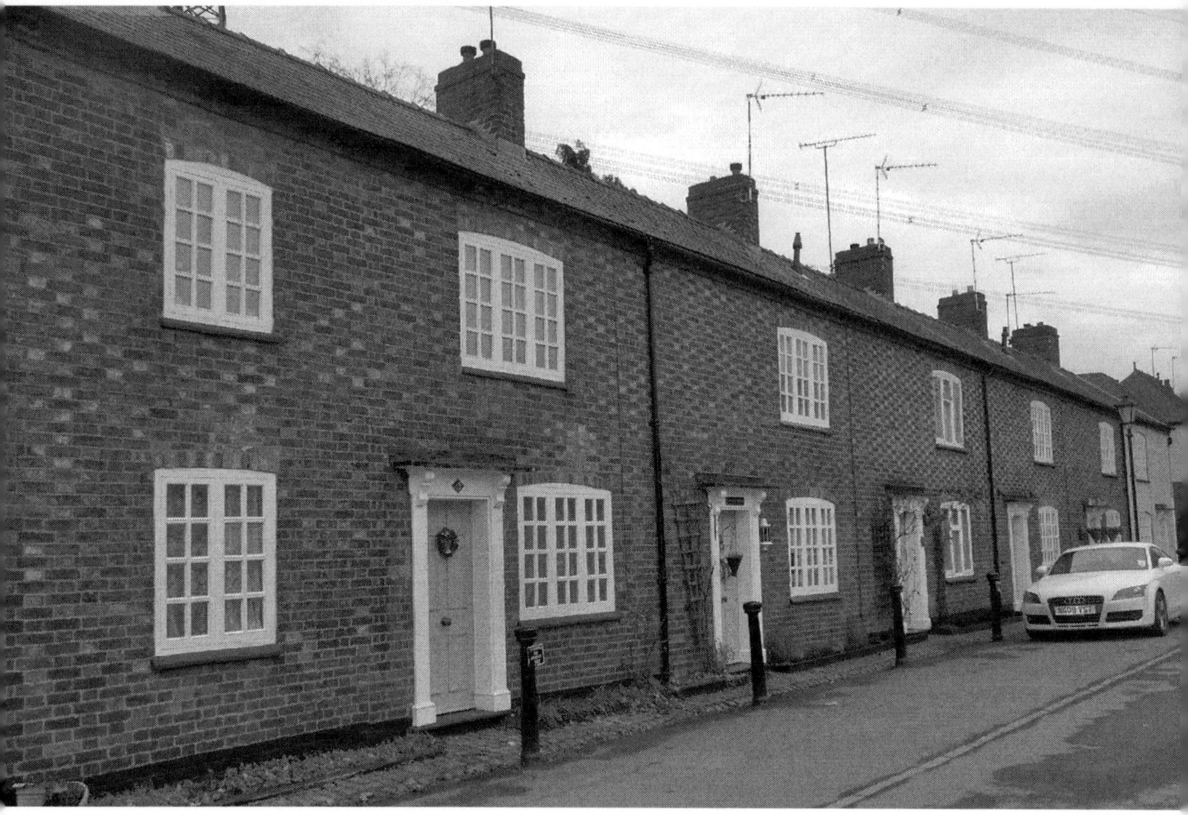

These dwellings at Hawkesbury Junction on the Coventry Canal are believed to be colliers' cottages associated with Victoria Colliery. They are in very good condition today and their diaper-pattern brickwork is very fine. (Anthony R. Case)

and future Labour Party Secretary Thomas Mann (1856–1941). He was born in a cottage on Victoria Farm in 1856 and, at the age of ten, took a job as a trapper at the colliery, where he worked until the colliery's closure in 1870.

JAMES WAGSTAFF

Builder and brick-maker of Nuneaton (may also have been a land valuer and auctioneer – active in the period 1835–41).

JOHN WILLIAMS

Manganese mining at Hartshill started in the early part of the nineteenth century, when black manganese nodules were found near to Hartshill Castle around about 1816. Manganese was part of a cocktail of components then recently discovered as an ingredient in a bleaching agent. Supplies of manganese in England at that time were very scarce, and it seems the manganese found at Hartshill was high-quality ore. How John Williams of Scorrier (near Truro), said to be the most extensive mine adventurer in Cornwall, became involved we may never know. But he did. The other manganese deposits then known were in Cornwall, Devon and Scotland, so having a deposit in the centre of England was a great commercial advantage. John Williams obtained a licence to work the land in August 1820. A crushing plant was built at Tuttle Hill and mining continued for about twenty years before another fourteen-year lease was signed in 1841. In 1845, the *Nuneaton Diarist* reported:

> The Manganese Mines which had been worked so extensively here and at Hartshill by the Cornish Mining Company for several years past have been bro't to a close. The Mines having become exhausted to a degree not to pay the working …

It seems some ore was still obtained for a few years thereafter, but all the economically workable material was worked out by the 1850s. There is still an adit into the hillside where manganese was obtained.

SAMUEL WOOD

This brickyard was located in Heath End, Nuneaton, and was operating in 1808. Later, business was carried on by George Sidwell, who subsequently went to America to sell his brick-making patent. The brickyard existed on land rented from the Harpur family in 1836.

WYKEN COLLIERY CO.

Lessees:

- Sir Thomas White's Charity, administered by Coventry Corporation (as was Hawkesbury and Craven (Sowe) collieries)
- Huntingdon Beaumont (1595)
- Sir Thomas Beaumont (1611)
- Earl of Dover (1636)
- Sir John Winter (1671)
- Evans & Broome
- James & Hall (1820)
- James Sibley Whitton (by 1856, and in 1881 he is listed as a farmer of 256 acres)

The colliery had its own farm attached to it, and was owned by Warwickshire Coal Co. Ltd from April 1912. It worked the Slate, Ell, Rider, and Two Yard seams.

The following are shafts associated with this colliery. Their operating dates are given where they are known:

- Pit No. 1
- Pit No. 2
- New Pit No. 1 (1856–64)
- New Pit No. 2 (1859–63)
- Walsgrave Main (1862–89)
- Basin Pit (1870–77), possibly the reopening of an earlier pit
- New Ironstone Mine (1870–77), worked sporadically
- Alexandra shaft (1880–1914)

Other known facts include the following:

- Newcomen engines were erected here in 1719, 1729, 1732 and 1734 respectively.
- Wyken Engine Pit was being worked in 1859.
- One shaft was retained as a pumping shaft for the Alexandra main colliery. This shaft was abandoned in 1933.
- A new set of shafts, referred to as Wyken Alexandra Colliery, was opened in 1886.
- There was a serious accident at the pit in 1889.
- The colliery employed 269 people underground and 132 on the surface.
- A brickworks was also operated at one time. It produced red bricks, drainpipes and tiles. Between June 1855 and March 1856, 407,850 bricks were produced here.
- Wyken merged with Craven Colliery; they closed together in 1928.

NOTABLE LOCAL BUILDINGS WITH LOCAL BRICKWORKS PROVENANCE

NUNEATON COUNCIL OFFICES

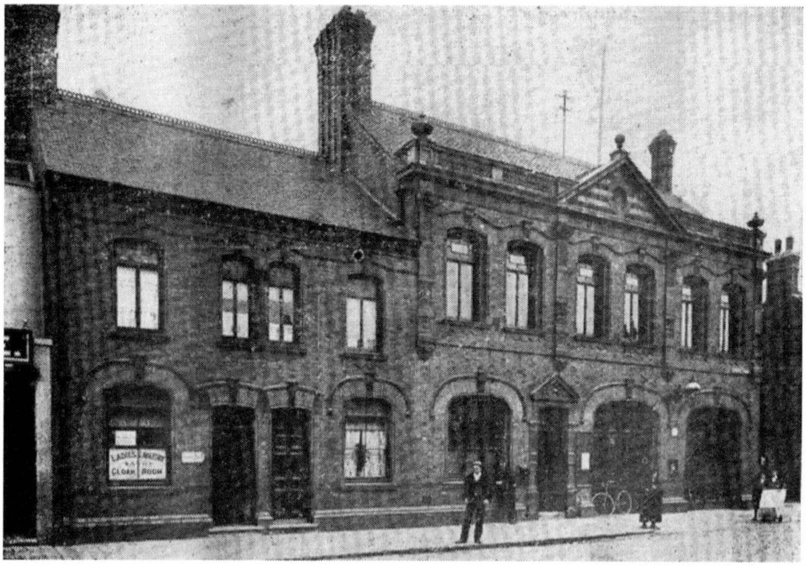

Nuneaton council offices, fire station and ladies' lavatory in Queens Road displayed some interesting terracotta features sourced from local brickyards. The window reveals and arches over the doorways were very finely ornamented. Finials on the corners of the main building completed the decoration. It was converted into the town library in 1933, when the new Council House offices opened in Coton Road. It is a pity it was demolished in the early 1960s and replaced with a modern concrete and brick rectangular box store complex and shopping arcade. (Author's collection)

Opposite above: Although the Hippodrome, which closed in 1956, was derelict in this view taken on 10 July 1968, the picture illustrates the fine-quality local terracotta work that originated in our local brickworks. It is most likely Stanley Brothers throughout, as the statue on the top came from there and was carefully taken down and preserved in Nuneaton Museum after the building was finally demolished, not long after this photo was taken. (Courtesy of Keith Draper)

THE HIPPODROME (FORMERLY PRINCE OF WALES) THEATRE

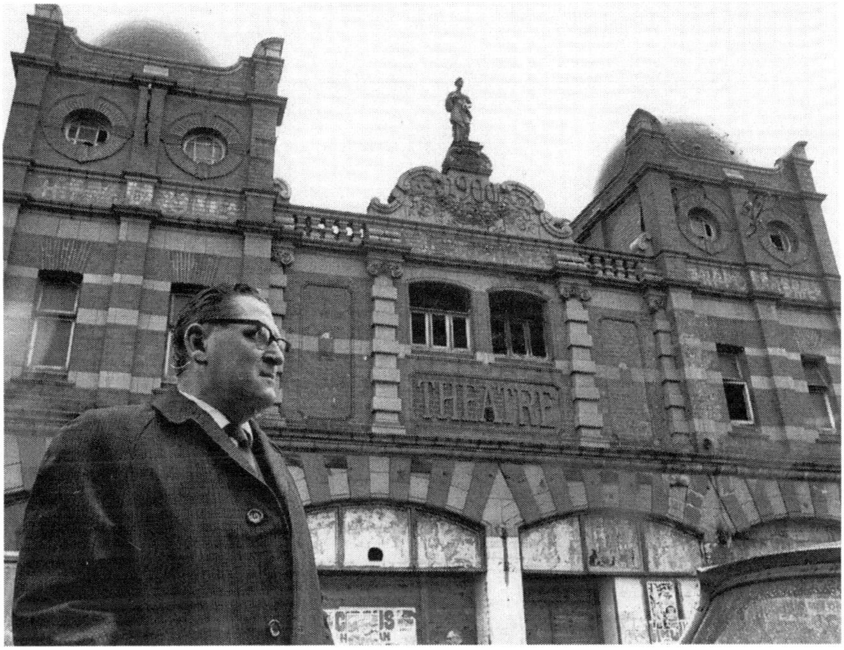

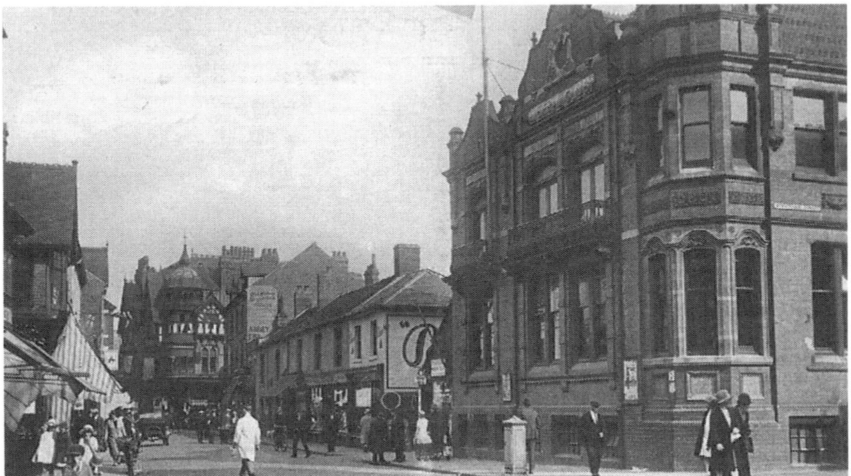

A view of Abbey Street towards the town centre illustrates two prominent local buildings utilising terracotta products from Stanley Brothers. The dominant building with a cupola in the middle distance was built at the cost of Reginald Stanley, the brickyard owner, in 1895. It incorporates many of his finest brick articles and was in effect a 3-D catalogue of the best Stanley's ware. Known as the Gate Temperance Hotel, it was centrally heated throughout on the most modern system for its day. There were spacious commercial rooms, a coffee lounge and a first-class restaurant. It closed as a hotel in 1908, but it survives today. On the right is Nuneaton Liberal Club (Reginald Stanley was a Methodist and a Liberal), which also survives and has fine terracotta features. The photograph was taken in the late 1920s. (Author's collection)

THE AUTHOR

Peter Lee has had a lifelong interest in his home town and the district of Nuneaton, Warwickshire, and is the foremost authority on the social and economic history of the area. He is Chairman of the Nuneaton Civic Society, the Nuneaton & North Warwickshire Family History Society, and co-organiser of the Nuneaton Local History Group. He writes a weekly column on local history for the *Nuneaton News*. He has contributed many articles to magazines and newspapers on local industrial archaeology and history over the years. He has written several books on the history of the area:

- *The Coventry & Nuneaton Railway*, Bedworth Echo, 1982 (co-authored with Mike Musson)
- *A History of Brewing in Warwickshire*, CAMRA, 1983 (with Fred Luckett and Ken Flint)
- *The Trent Valley Railway*, Trent Valley Publications, 1988
- *Images of England: Nuneaton*, Tempus Publishing, 2000
- *Images of England: Nuneaton, Vol. II*, Tempus Publishing, 2004
- *Railways of Nuneaton & Bedworth*, Tempus Publishing, 2005
- *Nuneaton Revisited*, Tempus Publishing, 2006
- *Nuneaton Through Time*, Amberley Publishing, 2009

Printed and bound by CPI Group (UK) Ltd, Croydon, CR0 4YY
27/08/2026
02205046-0005